AN INVESTIGATOR'S GUIDE TO SEA CRYPTIDS

An Investigator's Guide to Sea Cryptids

Dakota Frandsen

First Printing, 2025

Contents

1

Introduction to Sea Cryptids

The ocean's depths conceal a multitude of mysteries, some scientifically understood, others shrouded in legend. This book delves into the latter, exploring the realm of sea cryptids – creatures whose existence remains unproven, existing primarily in folklore, eyewitness accounts, and occasional ambiguous evidence. Defining what constitutes a sea cryptid requires a careful approach, acknowledging the line between genuine unknowns and embellished tales. A sea cryptid, for the purposes of this investigation, is defined as a purported marine animal whose existence lacks definitive scientific confirmation, yet for which credible evidence – however circumstantial – suggests a possibility beyond simple misidentification or fabrication.

This differs significantly from the purely mythical creatures found in numerous cultures' seafaring lore. While legends of krakens, sea serpents, and other monstrous beings have been passed down through generations, often serving as cautionary tales or explanations for unexplained phenomena, these are largely categorized as folklore. Sea cryptids, on the other hand, demand a more nuanced investigation. They occupy the space between established scientific understanding and the realm of the fantastical, requiring a blend of skepticism and open-mindedness in their examination.

The historical context of sea monster myths is vital in understanding the evolution of sea cryptid beliefs. Ancient mariners, facing perilous voyages and unpredictable ocean conditions, often attributed unexplained events to powerful, unseen forces. These encounters, often embellished through storytelling and cultural transmission, became ingrained in maritime traditions, shaping the collective imagination and contributing to a rich tapestry of sea monster legends. Consider the Norse kraken, a colossal cephalopod said to drag ships to their doom. Similar tales, although with varying characteristics, are found in many cultures, suggesting a potential underlying phenomenon, even if the specific details of the legend differ significantly. Is it possible that these accounts, viewed through a modern lens, point towards encounters with unusual or poorly understood marine creatures?

The geographical distribution of these legends also provides intriguing clues. Specific types of sea monsters are frequently associated with particular regions, suggesting potential links to the actual fauna inhabiting these areas. For example, the numerous accounts of sea serpents along the North American coast could reflect encounters with unusual deep-sea creatures occasionally surfacing near the shore. These geographic patterns lend themselves to further investigation, potentially pinpointing areas where the chances of encountering unidentified marine life are higher. Through mapping the locations of historical sightings and cross-referencing them with known marine habitats and migration patterns, investigators can begin to build a more complete picture.

However, it's crucial to consider the psychological factors influencing sightings and the subsequent interpretation of events. Factors like poor visibility, extreme weather conditions, and the power of suggestion can significantly skew perception and memory. The human brain, faced with ambiguity or uncertainty, often fills in the gaps with familiar patterns or established narratives, leading to misinterpretations of what is actually observed. This is not to dismiss eyewitness accounts outright, but rather to emphasize the importance of critically analyzing them. Interviews should be conducted with sensitivity and thoroughness, taking into account the individual's background, emotional state, and potential biases. Triangulating information from multiple, independent sources is crucial for verifying claims.

The role of mass media in shaping our perception of sea cryptids should not be overlooked. Sensationalized reporting and the proliferation of unverifiable images and videos online can significantly distort our understanding of the subject. The spread of misinformation, often amplified by social media, can easily lead to an inflated perception of cryptid activity. Therefore, responsible media consumption and a critical assessment of the information presented are crucial for a balanced approach to investigation. This is particularly true when analysing historical accounts, where artistic license, embellishments, and deliberate fabrications may heavily influence the details of the events described.

The psychology of belief also plays a crucial role in our understanding of sea cryptids. The enduring fascination with these creatures speaks to our inherent human curiosity about the unknown, the desire to unravel mysteries, and the appeal of the extraordinary. The search for sea cryptids touches on broader philosophical questions about the limits of human knowledge and our relationship with the natural world. Even in the absence of definitive proof, the persistent pursuit of these elusive creatures has driven technological advancements in ocean exploration and heightened our understanding of the vast biodiversity inhabiting Earth's oceans.

The exploration of sea cryptids, therefore, isn't merely a pursuit of the fantastical. It's a journey into the uncharted waters of scientific discovery, a blending of historical accounts, eyewitness testimony, and cutting-edge technology. It demands a rigorous methodology, employing the same scientific principles applied to the study of any other biological phenomenon, albeit with a greater degree of uncertainty and a higher tolerance for ambiguity. The definition of a sea cryptid, then, must constantly evolve as our understanding of the ocean's depths improves, as new technologies are developed, and as we learn to disentangle fact from fiction within the rich tapestry of seafaring lore. This process requires a careful balancing act between critical analysis and open-mindedness, a combination crucial for navigating the murky waters of cryptozoological investigation. The quest, as we will see, is ongoing, challenging, and potentially profoundly illuminating regarding the vast, largely unexplored ecosystems of our planet's oceans.

The tapestry of sea monster lore is woven from threads of fear, wonder, and the insatiable human curiosity about the unknown. For millennia, sailors and coastal communities have spun yarns of colossal creatures lurking beneath the waves, tales passed down through generations, evolving and morphing with each retelling. These narratives, far from being mere whimsical fabrications, offer a fascinating glimpse into humanity's relationship with the ocean, its anxieties, and its attempts to understand the mysteries it holds.

Ancient civilizations, dependent on the sea for sustenance and trade, developed rich mythologies surrounding its monstrous inhabitants. The Odyssey, a cornerstone of Western literature, features Scylla and Charybdis, two terrifying sea monsters that epitomize the perilous nature of seafaring. Scylla, a six-headed creature, snatched sailors from their ships, while Charybdis, a monstrous whirlpool, threatened to engulf entire vessels. These weren't simply random creations; they represented the real and ever-present dangers faced by sailors – treacherous currents, jagged rocks, and the unpredictable nature of the ocean itself. The legends served as cautionary tales, reinforcing respect for the sea's power and reminding mariners of their vulnerability.

Similarly, Norse mythology abounds with tales of sea monsters, most notably the kraken. This colossal cephalopod, capable of dragging ships to the depths, became a symbol of the ocean's unpredictable wrath. The kraken's image, often depicted as a gargantuan octopus or squid, was likely influenced by encounters with giant squids, although the legendary accounts greatly exaggerate their size and capabilities. The tales of the kraken, however, were not limited to the Norse; similar legends of enormous sea creatures capable of sinking ships appeared in the folklore of many maritime cultures. This suggests that the underlying fear and wonder inspired by these accounts stemmed from shared experiences and a collective understanding of the ocean's potential for devastation.

Moving beyond the ancient world, medieval and Renaissance bestiaries, illustrated manuscripts detailing real and mythical creatures, frequently included depictions of sea monsters. These often-fantastical portrayals were a blend of actual marine life—somewhat distorted and exaggerated—and entirely imagined creatures, a reflection of the limited scientific understanding of the time. The artistic depictions, while sometimes whimsical or grotesque, served to codify and disseminate the prevalent beliefs about the ocean's hidden inhabitants. These bestiaries reveal a striking blend of fact and fiction, showcasing the gradual process by which sea monster lore evolved, blending eyewitness accounts (however unreliable) with existing myths and superstitions.

The Age of Exploration brought with it a new wave of sea monster encounters, or at least, accounts of encounters. As ships ventured further into uncharted waters, so did the tales of the creatures they purportedly encountered. While many of these were undoubtedly embellished, some accounts present intriguing possibilities. Consider the numerous descriptions of sea serpents, long, serpentine creatures seen by sailors across the globe. While often dismissed as misidentifications of basking sharks or other large marine animals, the consistency of these reports across vastly different regions hints at something more complex. Could some of these accounts relate to unknown species, creatures that inhabit the ocean depths and rarely, if ever, surface?

The 19th and 20th centuries saw the rise of scientific investigation, leading to a more skeptical approach to sea monster stories. Naturalists and marine biologists sought to explain away reported sightings through rational explanations—misidentifications, optical illusions, or outright fabrications. However, the persistence of such reports, coupled with the advancement of underwater exploration technology, reignited interest in these elusive creatures. Sonar technology, capable of detecting objects underwater, occasionally registered unusual readings that couldn't be easily explained by known marine life, adding fuel to the sea monster debate.

The advent of photography and, later, videography, promised to provide irrefutable evidence. Numerous photos and videos have been offered as proof of sea cryptids, but many have been debunked as hoaxes, blurry images, or misinterpretations of familiar objects or animals. The challenge lies in separating genuine, unexplained sightings from the deliberate fabrications and misrepresentations that are often part of the narrative. Analyzing such evidence requires careful consideration of image quality, contextual information, and the potential for manipulation.

The proliferation of the internet and social media has amplified both genuine reports and hoaxes, blurring the lines further. Unverified images and videos of purported sea monsters rapidly spread, often without adequate scrutiny or investigation. This information overload makes it challenging to separate credible evidence from misinformation, highlighting the need for careful evaluation of sources and a critical approach to online content.

The study of sea monster lore, therefore, is not simply a matter of dismissing ancient myths or debunking modern-day hoaxes. It requires a nuanced approach that acknowledges the historical context of these legends, the psychological factors influencing eyewitness accounts, and the limitations of our understanding of the ocean's vast and largely unexplored depths. By combining historical analysis, critical evaluation of evidence, and a deep understanding of marine biology, we can hope to separate fact from fiction, to understand the cultural significance of sea monster lore, and potentially, to uncover some of the mysteries hidden beneath the waves.

The journey of investigating sea cryptids demands a constant reassessment of evidence, a blending of scientific rigor and open-mindedness, and a deep respect for the immense power and mystery of the ocean itself. The quest continues, promising to unravel more secrets with each investigation, each technological advancement, and each careful analysis of historical accounts. The ocean, vast and largely unknown, continues to inspire awe, fear, and a driving human curiosity that propels our ongoing exploration of its mysterious depths and its enigmatic inhabitants. The pursuit, far from being an exercise in chasing fantasy, is a reflection of our desire to understand the world around us, even in its most obscure corners. It is a testament to humanity's persistent questioning and our never-ending search for answers.

The allure of sea monsters transcends cultural boundaries and historical periods. While countless tales pepper maritime history, certain cases stand out due to their longevity, the volume of eyewitness testimony, or the intriguing nature of the evidence (or lack thereof). These "famous" cases, often shrouded in controversy and speculation, provide excellent case studies for understanding the complexities of cryptozoological investigation. Let's delve into some of the most prominent examples, examining the evidence, the methodologies employed (or lacking), and the inherent biases that color our interpretations.

The Loch Ness Monster, perhaps the most iconic sea cryptid, needs little introduction. Nestled in the murky depths of Scotland's Loch Ness, this elusive creature, often depicted as a plesiosaur-like animal, has captivated the world for decades. Numerous photographs, sonar readings, and eyewitness accounts have fueled the legend, yet conclusive proof remains elusive. Many images have been debunked as hoaxes or misidentifications of logs, boats, or even waterfowl. Sonar readings, while sometimes intriguing, are often difficult to interpret definitively, particularly in the challenging acoustic environment of a large, deep lake.

The methodological challenges in investigating the Loch Ness Monster are considerable. The sheer size of the loch, its depth, and the variable water conditions make systematic searches incredibly difficult. Furthermore, the reliance on eyewitness accounts, often collected decades after the alleged sighting, introduces significant potential for distortion and embellishment through memory lapses and the influence of pre-existing beliefs. The lack of standardized data collection procedures across different investigations further complicates the analysis. While the possibility of an unknown species cannot be entirely dismissed, a comprehensive evaluation of the existing evidence suggests the need for far more rigorous and systematic investigation before any definitive conclusions can be drawn. The enthusiasm surrounding the Loch Ness Monster, while understandable, often overshadows the importance of employing rigorous scientific methodologies.

Moving across the Atlantic, the Gloucester Sea Serpent offers another compelling case. Reported sightings of this creature, described as a long, serpentine animal, stretch back centuries along the New England coast. These sightings, often accompanied by detailed descriptions and even sketches, have been attributed to various large marine animals, including basking sharks and oarfish. However, some accounts defy easy explanation, leaving room for speculation. The lack of consistent physical evidence—such as clear photographs or video footage—remains a significant obstacle to definitive identification. Moreover, the inconsistencies in descriptions, likely influenced by the observer's perspective and pre-existing expectations, further complicate the analysis.

The Gloucester Sea Serpent case highlights the limitations of historical accounts. Eyewitness testimonies, especially those relayed over generations, can be unreliable due to memory distortion, the tendency to embellish narratives, and the influence of cultural context. The early depictions of the creature, often rendered in artistic representations, were necessarily influenced by the artistic style and the preconceived notions about sea monsters prevalent at the time. Furthermore, the absence of modern scientific tools and techniques in earlier investigations makes it nearly impossible to scrutinize the original accounts using contemporary standards of evidence evaluation. While we might never resolve the mystery of the Gloucester Sea Serpent definitively, analyzing the historical accounts allows us to better understand the evolution of the legend and the challenges of interpreting historical data in cryptozoological investigations.

Numerous other cases of purported sea monsters deserve attention, though they might lack the sustained media attention afforded to the Loch Ness Monster or the Gloucester Sea Serpent. Consider the numerous accounts of giant squid, some substantiated by recovered specimens or visual encounters. While giant squid are scientifically recognized, the sheer size they attain and their elusive nature fuel speculation about even larger, yet-to-be-discovered species. Similarly, the reported sightings of sea serpents in various parts of the world, though often attributed to misidentifications, continue to inspire debate and intrigue. The analysis of these accounts highlights the need for a nuanced approach, combining an understanding of known marine biology with a critical evaluation of eyewitness testimonies and any supporting evidence.

Many incidents involve misidentification – mistaking unusual behaviors of known species, unusual formations of marine plants or debris for something more fantastical. Large basking sharks, with their size and the way they sometimes lie near the surface, have been mistaken for sea serpents. Schools of fish can create visually deceptive patterns that could appear as something larger and more monstrous from a distance. The unpredictable and often challenging conditions of the ocean also add to the ambiguity – wave action, poor visibility, and the inherent difficulties of observing something vast in a very large environment can distort perspectives.

Beyond individual cases, a broader examination of methodological flaws in historical sea monster investigations is crucial. Early accounts often lacked the rigor of modern scientific methods, relying primarily on subjective eyewitness accounts with no established standards for data collection, analysis, or validation. The absence of standardized observation protocols or comparative analysis with known marine life often led to misinterpretations or exaggerations. Moreover, cultural biases and the influence of pre-existing beliefs inevitably shaped how sightings were reported and interpreted, potentially leading to the amplification of certain aspects while downplaying others.

The advent of modern technology, including underwater drones, sonar systems, and advanced photographic and video equipment, has revolutionized underwater exploration. Yet, even with these advancements, the vastness and complexity of the ocean present significant challenges. The deep sea remains largely unexplored, harboring numerous unknown species, and existing technologies don't guarantee the detection of elusive cryptids. The interpretation of sonar readings, for instance, can be ambiguous, requiring careful consideration of environmental factors and the potential for false positives. Similarly, blurry photos or short video clips, often the only evidence available, can be difficult to interpret definitively, leading to uncertainty and speculation.

Despite the challenges, the investigation of sea cryptids offers valuable insights into both marine biology and the history of human interaction with the ocean. By critically analyzing historical accounts and employing rigorous scientific methods, we can develop a deeper understanding of the complexities of eyewitness testimony, the potential for misidentification, and the limitations of our current understanding of the marine world. The pursuit of these elusive creatures, while ultimately reliant on evidence-based conclusions, also serves as a catalyst for scientific discovery and a reminder of the vast unknowns that still lie beneath the ocean's surface. The ongoing quest for these elusive creatures may not lead to their definitive discovery, but it constantly pushes the boundaries of scientific exploration and our understanding of the ocean's depths. The enduring fascination with sea monsters is, therefore, not simply a matter of pursuing fantasy, but of engaging in a profound and ongoing exploration of the unknown.

The pursuit of sea cryptids, while seemingly a venture into the realm of fantasy, necessitates a rigorous scientific approach. This isn't simply about debunking myths; it's about applying the same principles of scientific inquiry used to unravel the mysteries of any other unknown phenomenon. The core of this approach rests on several pillars: meticulous methodology, evidence-based reasoning, and a balanced combination of open-mindedness and healthy skepticism. Without these, the entire endeavor risks becoming nothing more than a collection of anecdotal stories, far removed from the realm of scientific investigation.

The scientific method provides the crucial framework. It begins with observation – not just casual sightings but systematic, repeated observations documented with precision. This involves detailed descriptions of the alleged cryptid, including size, shape, coloration, behavior, and the environment in which it was observed. Accurate record-keeping is paramount, employing standardized data collection forms to minimize bias and ensure consistency across multiple observations. Photographs and videos, while often blurry or inconclusive, should be collected and analyzed with the understanding that their limitations need careful consideration. The resolution, lighting conditions, and the distance from the subject all impact the reliability of the visual evidence.

Beyond visual data, other forms of evidence should be sought. Sonar readings, for instance, can provide indirect evidence of large objects in the water column. However, interpreting sonar data requires expertise and an understanding of potential artifacts and sources of error. Seabed sediments, water samples, or even biological traces – skin, scales, or other organic matter – could offer tangible evidence of an unknown creature. Gathering and analyzing this diverse range of evidence necessitates collaboration with specialists from various fields, including marine biologists, oceanographers, physicists, and chemists. This interdisciplinary approach expands the scope of investigation, ensuring a more comprehensive understanding of the potential evidence.

Crucially, the scientific method demands rigorous hypothesis testing. Before concluding that a sea cryptid exists, we must systematically eliminate alternative explanations. This often involves considering the possibility of misidentification – mistaking known marine life for something unknown. An oarfish, with its extraordinary length and serpentine form, could easily be mistaken for a sea serpent. Similarly, a school of fish, viewed from a distance, can create a visually deceptive pattern that might be misinterpreted as a larger, singular creature. The interplay of light, water clarity, and the observer's perspective can all introduce significant distortion.

Therefore, a crucial step in the process involves a thorough review of the biological diversity of the region where the alleged sightings occur. Understanding the known fauna is critical to distinguishing between a genuinely unknown species and a misidentified, known one. Comparing the observed characteristics to those of known marine animals – from common species to rarer ones – is essential for discerning any significant differences. Accessing and examining relevant databases of marine life, consulting taxonomic literature, and, if necessary, consulting specialists in marine taxonomy are vital steps in this process.

The analysis of eyewitness accounts necessitates a critical approach. Human memory is fallible; recollections can be distorted by time, cultural influences, and preconceived notions. The reliability of an eyewitness account is directly linked to the circumstances of the observation: distance from the subject, visibility, ambient lighting, the observer's state of mind, and their familiarity with marine life. To minimize bias, investigators should employ standardized interview techniques, ensuring that questions are unbiased and do not lead the witness toward a specific answer. The goal is to gather accurate and detailed descriptions, not to confirm preconceived notions.

Furthermore, the historical context surrounding the alleged sightings is crucial. Sea monster legends often have deep cultural roots, influenced by myths and folklore passed down through generations. Tracing the evolution of these legends, documenting the changes in descriptions over time, and examining the cultural context in which they arise are important steps in understanding the history of the belief. This historical analysis helps identify potential biases embedded within the narrative and distinguishes between credible observations and fantastical elaborations.

While rigorous skepticism is essential to avoid drawing premature conclusions, an equally crucial aspect is maintaining an open mind. The vastness and unexplored nature of the ocean's depths mean that the possibility of undiscovered species remains very real. Dismissing all unusual sightings without thorough investigation is not only unscientific but also hampers the discovery of genuinely new species. The balance between skeptical scrutiny and open-mindedness is the very essence of a robust scientific approach.

The application of modern technology greatly enhances the possibility of a successful investigation. Advanced sonar systems, remotely operated vehicles (ROVs), and autonomous underwater vehicles (AUVs) provide capabilities far exceeding those available to previous generations of cryptozoologists. High-resolution underwater cameras can capture detailed images and videos, offering clearer visual evidence than ever before. Genetic analysis of any collected biological samples can definitively identify or potentially identify an unknown species. The use of these technologies helps minimize the reliance on subjective eyewitness accounts, shifting the focus towards objective, quantifiable data.

However, even with modern technology, limitations remain. The challenges posed by the immense size and the diverse environments of the ocean are considerable. Deep-sea exploration remains expensive and logistically complex. The limitations of current technology might prevent the detection of cryptic species that possess evasive strategies or inhabit remote, inaccessible environments.

In conclusion, the investigation of sea cryptids demands a scientific approach that is rigorous, evidence-based, and balanced between healthy skepticism and open-mindedness. This involves a thorough investigation of known species, a critical analysis of eyewitness accounts, the application of modern technologies, and rigorous adherence to the scientific method. While definitive proof may remain elusive for many cases, the rigorous application of these principles ensures that the quest for sea cryptids remains a valuable exercise in scientific exploration and a catalyst for the discovery of new knowledge about our oceans. The pursuit continues to illuminate not only the potential existence of unknown creatures, but the profound mysteries that still remain within the vast, unexplored depths of our planet.

The pursuit of sea cryptids, while driven by scientific curiosity, must be undertaken with a profound sense of ethical responsibility. The ocean, a vast and complex ecosystem, is not simply a hunting ground for the elusive; it is a delicate web of life deserving of our utmost respect. Irresponsible investigative practices can inflict irreparable harm on this delicate balance, undermining the very environment we seek to understand. Therefore, ethical considerations form the bedrock of any legitimate cryptozoological research in the marine environment.

Central to this ethical framework is the principle of minimizing environmental impact. Deep-sea exploration, particularly when involving advanced technology like ROVs and AUVs, carries the potential for disruption to marine habitats. These vehicles, while invaluable tools for investigation, can inadvertently damage fragile ecosystems, disturbing delicate benthic communities or disturbing sensitive marine life. Careful planning, including thorough environmental impact assessments prior to any expedition, is paramount. Investigators must meticulously map the planned route, identify areas of high ecological sensitivity, and establish strict operational protocols to avoid harming marine life or damaging sensitive habitats. This might involve limiting the duration of submersible operations in particularly vulnerable zones, maintaining a safe distance from known breeding grounds or migratory routes, and utilizing non-invasive sampling techniques whenever feasible. The selection of research sites must also consider the potential disruption to existing marine research projects and initiatives.

Data collection itself necessitates ethical considerations. The acquisition of biological samples, even for the purpose of genetic analysis, must be conducted responsibly and with respect for marine biodiversity. Any sampling must be limited to what is absolutely necessary for the investigation and must adhere to all relevant permits and regulations. The use of non-lethal sampling methods is highly preferable, with minimal disruption to individual organisms and the ecosystem as a whole. Wherever possible, researchers should opt for non-invasive techniques such as underwater photography or video recording, supplemented by environmental DNA (eDNA) analysis. eDNA sampling, which involves extracting DNA from water samples, allows for the detection of species without requiring the capture or handling of organisms. This minimizes disturbance to the marine environment and avoids the potential for stress or injury to marine animals.

The responsible handling and storage of any collected samples are crucial. Biological material must be preserved using appropriate methods, minimizing degradation and ensuring the integrity of any subsequent analysis. Furthermore, researchers must adhere to stringent protocols for the ethical treatment of marine animals, particularly in cases where encounters with potentially sensitive species occur. Any observed marine life, regardless of identification, must be treated with care and respect. The avoidance of harassment, disturbance, or any action that could cause unnecessary stress or injury is non-negotiable.

Another crucial aspect of ethical cryptozoological research is the responsible dissemination of information. The spread of misinformation can have serious consequences, fostering unrealistic expectations, generating unwarranted fear, or leading to irresponsible actions that threaten marine ecosystems. Researchers must avoid sensationalism and hype, presenting findings in a factual and measured manner, acknowledging uncertainties and limitations where appropriate. The publication of research findings should follow established scientific peer-review processes, ensuring the rigorous scrutiny and validation of data before it reaches a wider audience. Transparency in methods and data analysis is essential, allowing other scientists to critically evaluate findings and replicate the investigation if necessary.

Moreover, the ethical researcher should be mindful of the potential for the misuse of findings. Claims of previously unknown marine creatures can provoke a variety of reactions, from excitement and wonder to fear and speculation. The researcher should always strive to present the information carefully, avoiding sensationalist language that may lead to public alarm or ill-informed actions that could affect marine wildlife or marine ecosystems. Unfounded claims can also lead to an erosion of trust in scientific endeavors, diminishing the credibility of genuine cryptozoological research and its contribution to oceanographic knowledge.

The investigation of sea cryptids is inherently intertwined with public engagement. However, it's vital to ensure that public interest is not exploited for personal gain or to sensationalize unfounded claims. Researchers should actively educate the public about the scientific methods employed, emphasizing the importance of evidence-based reasoning and critical thinking. The goal is not merely to generate excitement but to foster a deeper understanding of the ocean and its inhabitants. Collaboration with conservation organizations and governmental agencies is crucial to ensure that research findings are used to support conservation efforts and enhance marine protection strategies.

Furthermore, considering the international implications of research concerning marine environments is essential. Many marine species cross international boundaries, and collaborative efforts between nations are crucial for managing these shared resources and protecting biodiversity. The principles of collaboration, sharing data, and respecting international laws are integral to ethical research in this area. This cross-border collaboration also reduces duplication of efforts and ensures the most comprehensive investigations possible.

Finally, the ethical dimension extends to the researcher's own conduct. Maintaining the highest standards of integrity, honesty, and transparency in all aspects of the research process is paramount. Researchers must ensure that their actions are driven by a genuine commitment to scientific inquiry and environmental stewardship, not by personal ambition or financial gain. This commitment to responsible research will not only produce valuable scientific data but will also foster a deeper understanding of the delicate marine ecosystem and protect its vulnerable inhabitants. The pursuit of sea cryptids should be a catalyst for discovery and conservation, not a source of environmental harm or misinformation. By embracing these ethical considerations, we can ensure that the search for these elusive creatures remains a responsible and sustainable endeavor, contributing meaningfully to our understanding of the ocean's depths and protecting the intricate marine environment for generations to come.

2

Understanding Marine Life

The careful study of common marine species is paramount for any serious cryptozoological investigation. Misidentification is a significant hurdle in the pursuit of elusive creatures; a creature initially believed to be a cryptid might simply be a poorly understood or rarely sighted known species. Establishing a robust understanding of common marine life, therefore, forms the essential groundwork for distinguishing between the known and the unknown. This involves more than just recognizing the superficial characteristics; it requires a deep understanding of their behavior, habitats, and variations within species.

Let's begin with the colossal squid (*Mesonychoteuthis hamiltoni*). Often shrouded in mystery due to its deep-sea habitat, the colossal squid is a master of camouflage, blending seamlessly with the dark abyssal plains. Its massive size and unusual anatomy — including a pair of enormous eyes, the largest in the animal kingdom, and powerful hooks on its tentacles — have led to countless tales of giant sea monsters. Understanding its natural behavior, however, reveals a creature far less menacing than folklore would suggest. Though capable of enormous power, the colossal squid is primarily a solitary, ambush predator, feeding on fish and other squid in the frigid depths. Its limited mobility and primarily benthic lifestyle contrast sharply with the agile, fast-moving monsters often described in seafaring legends. Distinguishing its behavior – the slow, deliberate hunting, lack of aggressive surface activity – from the often described hyper-aggressive behavior of legendary sea serpents, for instance, becomes crucial for accurate identification.

Giant squid (*Architeuthis dux*) frequently share a similar fate of misidentification. While their massive size can certainly be intimidating, their behavior differs significantly from the descriptions of monstrous kraken. Giant squid, like colossal squid, are deep-sea dwellers. They are opportunistic feeders, primarily consuming fish and smaller cephalopods, but their hunting techniques are largely passive. Their primary defense mechanism against predators like sperm whales involves jet propulsion and inking, not the dramatic, ship-sinking attacks often attributed to them in historical accounts. Careful examination of eyewitness accounts, considering factors such as lighting conditions, distance, and the observer's state of mind, is critical to separating fact from embellished legend. Furthermore, the highly variable coloration and size range within the species, combined with the often distorted views gained at night or in turbulent seas, further complicate accurate identification. Analyzing photographic and video evidence, considering factors like water clarity, camera distortion, and the potential for digital manipulation, plays a crucial role in verifying sightings.

Another frequently misidentified creature is the oarfish (*Regalecus glesne*). This massive, ribbon-like fish is the longest bony fish in the world, reaching lengths of up to 36 feet. Its unusual appearance—long, silver body, flowing fins, and a vibrant red crest—has fueled numerous myths and legends of serpent-like sea monsters. However, oarfish are generally docile, slow-moving creatures inhabiting deep waters. They are rarely seen at the surface, often appearing only when sick or dying or being brought to the surface by currents. Their fragile bodies, not adapted for active hunting or defense, lack the power and aggression often attributed to legendary sea monsters. Their usual behavior is solitary, preferring the deep waters and feeding primarily on small crustaceans and jellyfish, a diet far removed from the massive prey often attributed to sea serpents. Understanding their usual depth, their movements in the water column, and the circumstances in which they appear on the surface are critical to differentiating them from other cryptid possibilities.

Deep-sea anglerfish, with their bioluminescent lures and grotesque appearance, contribute to the repertoire of potential misidentifications. The anglerfish, with their disproportionately large heads, sharp teeth, and the characteristic lure used to attract prey, certainly look menacing. But their size is quite modest compared to mythical sea monsters. They are primarily ambush predators, relying on their lure to attract smaller fish within striking distance. Their behavior is rather sedentary; they typically stay in one location, waiting for prey to come to them. The very nature of their environment – the pitch-black, immense pressure of the deep sea – limits any direct interaction with boats or humans, making encounters extremely rare and limiting the possibility of them being involved in ship-sinking incidents. Understanding their habitat preference and hunting strategy helps separate their behavior and appearance from more mobile and aggressive legendary creatures.

The gulper eel (*Eurypharynx pelecanoides*), with its large, expandable mouth and elongated body, is another deep-sea dweller that could be mistaken for a strange cryptid. Its peculiar appearance, with its massive jaws and relatively small body, lends itself to fantastic interpretation. However, its behavior is not aggressive; the gulper eel is a slow-moving, relatively passive predator that feeds primarily on small crustaceans and other deep-sea organisms. It lacks the power or agility that would be expected from a creature capable of the destructive feats attributed to some sea monsters. The rarity of observations, coupled with its deep-sea habitat, makes accurate assessment and comparison to legendary accounts even more difficult. Yet, understanding its size and behavior, limited solely to the deep, helps separate it from narratives of surface encounters with larger, more imposing creatures.

Large jellyfish, such as the lion's mane jellyfish (*Cyanea capillata*), with their extensive tentacles and potentially massive bell, have also been misinterpreted as monstrous sea creatures. Their stinging tentacles are certainly capable of inflicting pain, and the sheer size of the largest lion's mane specimens can indeed be awe-inspiring. However, their behavior is passive; they drift with the currents, utilizing their tentacles to capture prey. Their lack of aggressive behavior and the ephemeral nature of their large size (these giants are often torn and fragmented, making them appear smaller and less menacing) differentiates them from the active predatory behavior implied in many sea monster legends. Understanding the life cycle and natural fragility of these gelatinous giants helps differentiate their appearance and behavior from the strength and aggression attributed to cryptids.

The study of these common species, coupled with a detailed understanding of their anatomy, behavior, habitat, and the potential for misidentification, serves as an essential stepping stone in the responsible investigation of sea cryptids. By diligently separating fact from fiction based on observational and scientific data, we can better understand the possibilities – and limitations – of what lurks beneath the waves. This rigorous approach, grounded in scientific principles and ethical considerations, is crucial not only for unraveling the mysteries of the deep but also for protecting the delicate balance of life in the ocean environment.

The deep sea, encompassing the vast expanse below 200 meters, represents the largest and least explored habitat on Earth. Its crushing pressures, perpetual darkness, and frigid temperatures create a unique environment teeming with bizarre and often unseen life. Understanding the ecosystems and inhabitants of this realm is crucial to any serious cryptozoological investigation, as many purported sea monsters are likely rooted in encounters with unusual deep-sea creatures. The abyssal zone, in particular, harbors a wealth of organisms adapted to survive in these extreme conditions, many of which are only recently becoming known to science.

One of the most striking adaptations found in deep-sea organisms is bioluminescence, the production and emission of light. Many creatures utilize bioluminescence for a variety of purposes, including attracting prey, luring mates, or deterring predators. Anglerfish, for example, possess a bioluminescent lure that dangles in front of their massive mouths, attracting unsuspecting smaller fish into their jaws. This light, often in a specific color and pattern, can create a striking visual, potentially leading to misinterpretations in low-light conditions. A fleeting glimpse of bioluminescence in turbulent waters could easily fuel tales of glowing sea monsters. Other creatures, such as certain species of squid and jellyfish, employ bioluminescence in more diffuse patterns, creating shimmering displays that may be misinterpreted from a distance or in poor visibility.

Hydrothermal vent ecosystems represent another remarkable example of deep-sea biodiversity. These vents, found along mid-ocean ridges, release superheated, chemically-rich water from the Earth's interior. Despite the extreme temperatures and toxic chemicals, these vents support vibrant communities of life, entirely independent of sunlight. Unique organisms, such as tube worms, giant clams, and chemosynthetic bacteria, thrive in this environment, utilizing the chemicals released from the vents as an energy source. These organisms often possess unusual morphologies and physiological adaptations that, if observed briefly or incompletely, could be readily misconstrued as cryptid creatures. The unique colors and shapes of vent organisms, coupled with the extreme environment and limited visibility, can fuel imagination and contribute to misidentification.

The cold seeps, similar to hydrothermal vents, are another area of significant deep-sea biodiversity. Cold seeps release methane and other hydrocarbons from the seafloor, providing an energy source for specialized chemosynthetic communities. These communities, like those found around hydrothermal vents, often feature unique and unusual organisms, adapted to survive in these chemically-rich environments. These unusual creatures may not resemble anything found in shallower waters, increasing the likelihood of misidentification if encountered by researchers, or, more likely, by fishermen working in areas with cold seeps, who might subsequently embellish a deep-sea sighting into a cryptid account.

The deep-sea floor itself is home to a diverse array of organisms, including various types of fish, invertebrates, and microbes. Many deep-sea fish possess unusual adaptations, such as large mouths, bioluminescent organs, and expandable stomachs, that allow them to survive in the food-scarce environment. Their often strange and unusual appearance combined with the limited knowledge of deep sea species makes it easy to mistake them for something else, contributing to cryptid narratives.

Studying these deep-sea ecosystems presents significant challenges. The immense pressure, darkness, and cold temperatures make direct observation difficult and expensive. Submersibles and remotely operated vehicles (ROVs) are often necessary to explore these environments, and even with these advanced technologies, exploration is limited by both technological capabilities and the sheer size of the deep sea. Furthermore, the sampling techniques required for deep-sea research can be highly disruptive to these fragile ecosystems, highlighting the importance of ethical considerations in any deep-sea exploration and research, especially when related to cryptozoology. The potential damage to the habitats, to the organisms themselves, and to the delicate balance of the deep sea ecosystems should always be a primary consideration.

The difficulty in directly observing deep-sea creatures often leads to reliance on indirect methods such as sonar, underwater photography, and video. Sonar, while useful for mapping the seafloor and detecting large objects, can sometimes produce ambiguous results, leading to misinterpretations of what is actually being observed. Underwater photography and video, while providing visual evidence, can be limited by the clarity of the water, the lighting conditions, and the distance from the object being observed. The technological limitations inherent in deep-sea research can lead to interpretations of blurry images and indistinct sonar pings as evidence of unknown creatures, when in fact, the observations simply lack clarity and resolution.

Another crucial factor in understanding the deep sea and its potential for cryptid sightings involves recognizing the limitations of historical accounts. Many sea monster legends have been passed down through generations, often embellished or distorted over time. It's important to carefully examine these accounts, considering the cultural context, the biases of the observers, and the potential for misinterpretations of naturally occurring phenomena. Many legendary descriptions lack precision, leading to broad interpretations which could fit many different sea creatures – and could even be entirely fictional.

Furthermore, the sheer size of the deep sea and the limited exploration undertaken means that countless unknown species likely remain undiscovered. Therefore, it's important to approach any claims of cryptid sightings with a combination of open-mindedness and rigorous scientific scrutiny. A careful evaluation of evidence, along with a robust understanding of known deep-sea organisms, is essential for distinguishing between established species and potentially undiscovered creatures. It's a balancing act of keeping an open mind about the unexplored while applying rigorous and critical scientific methods to evaluate the information that does arise.

The exploration of deep-sea ecosystems and their inhabitants is an ongoing process. As technology advances and more research is conducted, our understanding of these environments and the creatures that inhabit them will continue to improve. This ongoing process of research, however, underscores the importance of responsible investigation, respect for the delicate balance of these deep sea environments and the species within, and the critical need to distinguish between scientific inquiry and sensationalism in any cryptozoological exploration.

The deep sea is a fascinating and largely unexplored realm, holding countless mysteries. A careful and responsible approach to understanding its mysteries will not only unlock secrets about our planet but also safeguard its fragile ecosystems for future generations. The quest for understanding the deep sea, therefore, is a complex task demanding rigorous science, a commitment to ethical principles, and a healthy dose of skepticism – a combination that ultimately benefits both our pursuit of the unusual and the well-being of this unique and significant ecosystem.

The ocean's vastness holds a staggering array of life, much of it still unknown to science. While the iconic whales, dolphins, and sharks readily capture our attention, a hidden world of unusual and rarely seen creatures thrives beneath the waves, often inhabiting the deep ocean's unforgiving environments. These organisms, with their peculiar adaptations and often bizarre appearances, can easily be mistaken for legendary cryptids, fueling speculation and adding to the mystique surrounding these elusive beings. Understanding these lesser-known species is, therefore, critical to any serious investigation into purported sea monsters.

One group deserving particular attention is the deep-sea anglerfish. These bizarre creatures, found in the abyssal depths, are masters of camouflage and ambush predation. Their most striking feature is their bioluminescent lure, a modified dorsal fin spine that dangles enticingly in front of their enormous mouths. This lure, emitting a soft, alluring glow, attracts smaller fish, which are then swiftly engulfed by the anglerfish's gaping jaws. The stark contrast between the dark, featureless background of the deep sea and the bright lure could easily contribute to exaggerated or misinterpreted accounts of a mysterious light source, potentially fueling sea monster myths.

Furthermore, the anglerfish's unusual morphology—its large, disproportionate head, sharp teeth, and overall unsettling appearance—only adds to the potential for misidentification. Their varied sizes, ranging from a few centimeters to over a meter in length, further contribute to the diversity of sightings, which might then be incorrectly attributed to multiple cryptid species. Detailed illustrations and photographic comparisons, available in numerous scientific publications and online databases, can help differentiate these deep-sea denizens from more mysterious claims.

Moving beyond the anglerfish, the vast family of deep-sea squid presents another fascinating group. Many species possess unusual adaptations, such as bioluminescent organs, chromatophores (cells that rapidly change color), and remarkable camouflage abilities. Their ability to rapidly change their appearance, combined with their often elusive behavior, makes positive identification in the wild exceptionally difficult. Giant squid, in particular, have long been associated with sea monster lore. Their immense size, reaching lengths of up to 13 meters, and their elusive nature contribute to their legendary status.

However, scientific studies have shown that despite their size, giant squid are not inherently aggressive; most encounters are accidental and result from the squid's natural behavior rather than any intent to attack. The limited opportunities to observe giant squid in their natural habitat, combined with their elusive nature and their size which can easily lead to miscalculation, means many anecdotal stories and eyewitness accounts lack the accuracy needed for confident identification, and can easily be misinterpreted. Careful examination of photographic and video evidence, focusing on features such as tentacle morphology and body proportions, alongside the application of established taxonomic keys, can help mitigate this issue.

The ocean's depths also harbor numerous species of jellyfish, some of which possess equally striking characteristics. Certain deep-sea jellyfish species exhibit stunning bioluminescence, creating mesmerizing displays of light in the otherwise inky blackness. The vast range in sizes and forms among jellyfish, ranging from the relatively small to massive species several meters in diameter, can lead to considerable confusion when interpreting encounters, especially in low light or turbulent conditions. Their often translucent bodies, combined with their pulsating movements, might readily inspire accounts of strange, gelatinous sea creatures. High-quality photographic documentation, incorporating scale references and noting the specific location and conditions of the observation, are invaluable tools for accurate identification and distinguishing these species from any alleged cryptid phenomena.

Beyond invertebrates, the deep-sea fishes present a treasure trove of unusual and rarely observed species. Many possess adaptations suited to the extreme pressures, cold temperatures, and scarcity of food found in the abyssal depths. Some deep-sea fish possess large, gaping mouths, allowing them to effectively capture any prey that comes within range. Others have bioluminescent organs, utilized for attracting mates or prey. The often strange and unusual morphologies of these fishes, combined with the limited scientific knowledge about many of these species, easily lends itself to misinterpretations. Their appearance, amplified in the low-light conditions of the deep ocean, could easily contribute to the perpetuation of sea monster tales. Careful scrutiny of anatomical features, such as fin shapes, dentition, and body proportions, is essential for accurate identification. Moreover, relying on multiple lines of evidence, such as photographic and video recordings, along with sonar data (if applicable), can significantly improve the accuracy of identification.

Furthermore, the hydrothermal vent and cold seep ecosystems, often found deep below the surface, are teeming with life forms so utterly unique they challenge our conventional understanding of biology. These chemosynthetic communities, thriving on chemicals from the Earth's interior, harbor a diverse array of organisms, many of which display extraordinary adaptations and morphologies. Giant tube worms, with their feathery plumes, and massive clams are but two examples. These organisms, with their unusual appearance and the often extreme environments they inhabit, could readily fuel the imagination and lead to claims of unidentified sea creatures. The sheer uniqueness of these creatures, unfamiliar to most observers, readily lends itself to misidentification and exaggerated accounts. Scientific literature and research databases, containing detailed information and imagery of vent and seep organisms, are essential resources in these investigations.

Finally, it's imperative to acknowledge the limitations of historical accounts of sea monsters. Many such narratives have been passed down through generations, frequently embellished or distorted over time. Cultural context, the biases and perceptual limitations of the observers, and the propensity for human beings to interpret ambiguous information according to pre-existing beliefs and expectations all contribute to the reliability issues. It's crucial to critically evaluate these accounts, distinguishing between verifiable evidence and folklore. A careful and rigorous approach, combining scientific knowledge with a critical assessment of historical and anecdotal evidence, is essential for discerning between established species and genuinely unknown marine organisms.

In conclusion, the ocean's depths are replete with a surprising array of bizarre and rarely seen creatures. Their unusual adaptations, peculiar morphologies, and the challenging conditions of their habitats frequently contribute to misidentification and the perpetuation of sea monster myths. By meticulously studying these lesser-known marine species and understanding their unique characteristics, we can enhance our ability to distinguish them from truly unknown creatures, furthering our comprehension of the ocean's hidden wonders and potentially resolving some of the longstanding mysteries surrounding alleged cryptids. The combination of scientific rigor, technological advancement, and a deep understanding of the ocean's biodiversity is essential for progressing in this field and safeguarding the delicate balance of our marine environments. The journey into the heart of the ocean's mysteries is a testament to the continuous interplay between discovery and careful investigation.

The identification of marine species, a cornerstone of marine biology, becomes particularly crucial when investigating potential cryptids. While anecdotal accounts and blurry photographs often fuel speculation, rigorous scientific methods are necessary to differentiate established species from truly unknown organisms. This requires a multi-faceted approach, employing a range of techniques that go beyond simple visual identification. The accuracy of identification directly impacts the validity of any cryptid claim, making these techniques essential to a thorough investigation.

One of the fundamental techniques involves meticulous anatomical analysis. This process entails a detailed examination of the physical characteristics of a specimen, whether it's a live animal, a carcass, or even skeletal remains. For marine animals, specific features like fin shape and number, dentition (teeth structure and arrangement), body proportions, skin texture, coloration patterns, and the presence of unique structures such as bioluminescent organs are all carefully documented. Detailed measurements are crucial, establishing quantitative data that allows for comparisons with existing species descriptions. High-resolution photography and videography, along with detailed sketches, form an integral part of this process, providing permanent records for analysis and comparison.

Anatomical analysis extends beyond gross morphology. Microscopic examination of tissues and cellular structures can reveal additional identifying features. For instance, the microstructure of scales, the arrangement of muscle fibers, or the presence of specific types of cells can provide crucial taxonomic information, especially in cases where external morphology is ambiguous or damaged. This level of detail is often necessary when dealing with fragmented remains or highly decomposed specimens, circumstances often encountered in cryptid investigations.

Furthermore, the context of the discovery is crucial. The geographic location, depth of the water, and the surrounding ecosystem all provide valuable ecological data that can aid in species identification. Some species exhibit highly localized distributions, meaning their presence in an unusual location could immediately suggest a misidentification or a truly unique discovery. Analysis of the animal's diet, determined through stomach content analysis, can also be insightful, providing clues about the animal's niche within its ecosystem and thus narrowing down the possibilities.

Beyond anatomical observation, modern marine biology employs powerful molecular techniques, particularly DNA sequencing. A small tissue sample, even a trace amount of skin or mucus, is sufficient to extract DNA and compare it to extensive databases containing genetic information from known marine species. This method allows for precise identification even when dealing with partial specimens or those exhibiting unusual morphological features. DNA barcoding, a technique using specific genetic markers to rapidly identify species, has become increasingly valuable, accelerating the identification process and potentially revealing cryptic species or hybrids previously unknown to science. Phylogenetic analysis, comparing the genetic sequences of the unknown organism to those of related species, can provide further insights into evolutionary relationships and taxonomic classification. The advancement of next-generation sequencing technologies has dramatically increased the speed and efficiency of these analyses, enabling rapid identification and comparison even with minimal genetic material.

Ecological data analysis plays a vital role in corroborating identification. Analyzing the environment where the organism was discovered, observing its behavior, and studying its interactions with other organisms can provide strong supporting evidence for its classification. For example, the presence of specific parasites or symbiotic organisms can be highly indicative of a species' identity. Stable isotope analysis, measuring the relative abundance of different isotopes within an animal's tissues, can reveal details about its diet and trophic level within the ecosystem, further enhancing identification certainty. The combination of anatomical, molecular, and ecological data forms a robust approach, significantly enhancing the reliability of species identification, thereby allowing investigators to effectively rule out known species in the pursuit of potential cryptids.

However, it's crucial to acknowledge the limitations of these techniques. DNA degradation in decaying specimens can hinder molecular analysis. Similarly, in the case of highly unusual and potentially unique organisms, current databases might lack sufficient comparative material, leading to inconclusive results. Also, the interpretation of anatomical data can be subjective, especially when dealing with specimens that have been significantly altered by environmental factors or decomposition. In such cases, the expertise of experienced marine biologists and taxonomists is essential in critically evaluating the available evidence and drawing informed conclusions.

Further complications arise from the sheer diversity of marine life, much of it yet unexplored. Deep-sea environments, in particular, are largely unknown, harboring a vast array of species with unique adaptations and morphologies. Many deep-sea organisms have only been observed in limited contexts, leaving their biological understanding incomplete. This necessitates a cautious approach, acknowledging the possibility of encountering truly novel organisms while simultaneously maintaining a robust scientific methodology to prevent misinterpretations.

The analysis of photographic and video evidence often forms the initial phase of cryptid investigation. However, interpreting such evidence needs a cautious approach, recognizing the limitations of the imagery. Factors like poor lighting conditions, distance from the subject, water clarity, and the use of magnification can significantly distort the appearance of an organism, making accurate identification based solely on visual data unreliable. In many cases, such images can be ambiguous, leading to multiple possible interpretations. Combining visual evidence with other data points, such as ecological information and witness testimonies (if available and reliable), is crucial for a more comprehensive and accurate assessment.

It is also essential to acknowledge that even with advanced techniques, definitive identification isn't always possible. In situations involving incomplete or degraded specimens, or when dealing with truly novel organisms, some level of uncertainty will persist. This emphasizes the importance of transparency in reporting and the need to acknowledge the limitations of the available data. Well-documented investigations, meticulously outlining the methods used and clearly stating the uncertainties encountered, are far more valuable than speculative claims based on incomplete information.

In conclusion, the identification of marine organisms, especially when investigating potential cryptids, is a rigorous process that requires a multidisciplinary approach. Combining traditional anatomical analysis with cutting-edge molecular techniques, along with detailed ecological data, significantly improves the accuracy of species identification. However, researchers must remain aware of the limitations of available technologies and methodologies, always striving for transparency and rigorous reporting. The ultimate goal is not merely to identify known species but also to uncover and understand the full diversity of marine life, fostering a greater appreciation for the ocean's hidden wonders. This approach ensures responsible scientific investigation, fostering a balance between open-mindedness and critical thinking essential to unraveling the mysteries of the deep.

Misidentification in marine biology, especially within the context of cryptozoology, is a pervasive challenge. It's not simply a matter of mistaking a dolphin for a porpoise; the sources of error are far more subtle and multifaceted, demanding a rigorous approach to investigation. Understanding these sources is paramount to separating credible cryptid claims from instances of misidentification. Let's delve into the key factors contributing to these errors.

One major source stems from the inherent limitations of underwater observation. The marine environment presents considerable challenges to accurate visual identification. Poor visibility, often caused by turbidity, suspended sediment, or dense biological aggregations (like plankton blooms), severely restricts the clarity of underwater observations. Distance further exacerbates the problem. Even with advanced underwater camera systems, discerning fine details on a large animal at a considerable depth can be exceedingly difficult, resulting in blurred images and ambiguous shapes, fueling speculation and misidentification. The refraction of light underwater can also distort the apparent size, shape, and color of objects, adding another layer of complexity to visual assessments. In low-light conditions, common in deep-sea environments, color perception is compromised, leading to inaccurate estimations of an organism's true coloration. This is frequently compounded by the effects of bioluminescence, which can create dazzling and potentially misleading light displays.

Optical illusions, common in any field of observation, are particularly potent underwater. The interplay of light, shadows, and water currents can create deceptive patterns and shapes, leading observers to interpret ambiguous forms as something extraordinary. For example, large kelp forests or underwater rock formations, partially obscured by sediment or poor visibility, may be mistakenly perceived as large, unknown creatures. Similarly, schools of fish, particularly those with reflective scales or unusual swimming patterns, can easily be misidentified as a singular, gigantic organism from a distance. The shifting patterns of light and shadow can transform a familiar creature into something seemingly monstrous.

The psychological aspects of observation also play a significant role. Confirmation bias, the tendency to favor information confirming pre-existing beliefs, is a particularly insidious source of error. An observer expecting to see a sea monster might interpret ambiguous shapes and movements in a manner that supports that expectation, overlooking alternative explanations. This bias can be amplified by the excitement and adrenaline associated with a potential discovery, leading to hasty and inaccurate assessments. The phenomenon of pareidolia, the tendency to perceive familiar patterns in random stimuli (such as seeing faces in clouds), can further influence perception, transforming vague underwater shapes into recognizable—but completely inaccurate—forms. Fear, uncertainty, and the pressure of making a "discovery" can exacerbate these cognitive biases, leading to misinterpretations.

The limitations of reporting also significantly impact the reliability of observations. Eyewitness accounts, even when well-intentioned, can be profoundly unreliable. Memory is notoriously fallible, and the details of an observation can be distorted, exaggerated, or even entirely fabricated over time. The language used to describe an event can also introduce inaccuracies. Vague terminology, lack of precise measurements, and the influence of cultural narratives can all contribute to misinterpretations. For example, a description of a "large, serpentine creature" could encompass a wide range of possibilities, from a colossal squid to an aggregation of eels, to a floating piece of debris mimicking a serpentine form.

Photographic and video evidence, while seemingly more objective, also present challenges. The quality of the imagery significantly impacts the reliability of the identification. Low-resolution images, poor lighting conditions, and motion blur can all obscure crucial details, rendering accurate identification impossible. Furthermore, the angle of the photograph or video, the presence of other objects in the frame, and even the processing of the digital image can all contribute to misinterpretations. A grainy, distant photograph might be interpreted as evidence of a previously unknown creature when a closer examination, or additional photographic evidence, reveals it to be a known, if unusual, species or a natural phenomenon. The lack of scale in many underwater photographs further complicates identification. A smaller creature in the foreground may seem larger than it is in comparison to something in the background. The same image could be presented in two different ways: one where a small creature appears huge due to relative size, the other using a scale to show its true size. Digital manipulation, intentional or accidental, further obscures the integrity of photographic evidence.

The lack of standardized reporting procedures in cryptozoological investigations exacerbates the problems of misidentification. Inconsistencies in data collection, the absence of detailed measurements, and the lack of rigorous methodological rigor often render evidence ambiguous and unreliable. In contrast, rigorous scientific investigations employ standardized methods, including detailed photographic documentation, precise measurements, sample collection, and meticulous record-keeping, significantly minimizing the potential for errors. The critical difference lies in the controlled and structured nature of scientific investigations, which reduces biases and enhances the credibility of findings.

Finally, the sheer diversity of marine life remains a significant hurdle. Many species are poorly understood, with limited scientific knowledge of their morphology, distribution, and behavior. This lack of baseline information makes it inherently difficult to distinguish known, but rarely seen, species from truly novel organisms. The deep sea, in particular, remains largely unexplored, harboring a wealth of undiscovered species with unusual adaptations. Until a comprehensive catalog of marine life exists, misidentification will remain a persistent challenge in any investigation involving unusual marine sightings. The ocean's vastness and the inherent difficulties of underwater observation mean that many discoveries are made based on a limited sample size, leading to uncertainties and the potential for misidentification. This highlights the need for continuous exploration and research to better understand the vast diversity of marine life.

In conclusion, avoiding misidentification in marine cryptid investigations requires a multi-pronged approach. It's crucial to acknowledge the limitations of underwater observation, understand the psychological biases that can affect perception, and employ robust reporting techniques that emphasize precise data collection and rigorous analysis of all available evidence. Furthermore, a thorough understanding of known marine life, combined with a healthy skepticism and a commitment to sound scientific methodology, are essential tools for separating credible evidence from misidentifications. Only through such careful and systematic investigation can we hope to uncover the true secrets of the deep, and more importantly, differentiate actual discovery from mere illusion.

3

Investigating Eyewitness Accounts

Analyzing eyewitness accounts in the realm of cryptozoology, particularly concerning potential sea cryptids, necessitates a nuanced and critical approach. While anecdotal evidence can provide intriguing leads, its inherent limitations must be carefully considered before drawing any conclusions. The pursuit of truth demands a rigorous evaluation, moving beyond simple acceptance or dismissal. Several key strategies are crucial in this process.

First, corroboration is paramount. A single eyewitness account, no matter how vivid or detailed, lacks the weight of multiple, independent accounts supporting the same core observations. Discrepancies between different accounts can be equally informative, pointing towards potential misinterpretations or biases. For instance, if multiple witnesses describe a large, serpentine creature, but their descriptions of size, color, and behavior differ significantly, it suggests a need for further investigation, rather than immediate acceptance of a new species. The geographical location of the sightings, the time of year, and environmental conditions should all be considered during corroboration. Multiple sightings in the same region, under similar conditions, strengthen the plausibility of a genuine phenomenon.

Determining the reliability of each witness is crucial. This involves evaluating their background, experience, and potential biases. A seasoned mariner with years of experience at sea would likely have a different level of observational skill and judgment than a casual beachgoer. Their familiarity with marine life, including known species and unusual behaviors, will impact the credibility of their claims. Moreover, the witness's emotional state during the observation must be considered. Fear, excitement, or a desire for attention can all significantly distort memory and perception. A calm and methodical approach to questioning witnesses is crucial. Open-ended questions that encourage detailed descriptions, without leading the witness, will provide more valuable insights. It's important to note here that witness testimony should never be taken at face value. Even the most credible-sounding eyewitness may be subject to misinterpretations or subconscious biases.

Addressing potential biases is an essential component of a thorough analysis. Confirmation bias, as previously discussed, can significantly influence perceptions and recollections. A witness strongly believing in the existence of sea monsters might interpret ambiguous observations to fit their preconceived notions, overlooking alternative, simpler explanations. Similarly, the influence of cultural narratives and folklore can shape descriptions, leading to exaggerated or embellished accounts. A sighting described as a "Leviathan" might be a whale, a large school of fish, or even a cleverly shaped piece of driftwood, depending on the cultural lens through which it is interpreted. It is crucial to isolate the core observation from the embellishments or biases added later.

Another critical aspect is evaluating the witness's level of detail. Vague descriptions are less reliable than those with precise measurements and contextual information. A statement like "I saw a huge creature in the water" offers limited information, while "I saw a creature approximately 30 meters long, dark grey in color, with multiple appendages, approximately 1 kilometer off the coast near the kelp forest at approximately 2 pm on July 15th" is considerably more informative and verifiable. The more specific and measurable the description, the more easily it can be evaluated against existing knowledge of marine life.

The use of external corroborating evidence is also crucial. Physical evidence, such as photographs or videos, can complement eyewitness accounts, providing tangible support or raising new questions. However, as previously noted, the quality and context of such evidence are paramount. Low-resolution images, unclear videos, or images that lack scale can be difficult to interpret accurately. Furthermore, the possibility of image manipulation or misinterpretation must be considered. Sonar data, if available, can provide valuable insights into the size and shape of underwater objects, corroborating or contradicting eyewitness testimony. However, sonar data interpretation also requires careful evaluation, recognizing the limitations of technology and the potential for false positives.

Geographical and historical research can be vital in analyzing eyewitness accounts. Examining historical records, local legends, and maritime charts can reveal patterns of similar sightings over time, identifying potential hotspots or recurrent observations. Understanding the local ecology and known marine species is essential to distinguishing genuine anomalies from familiar creatures behaving unusually. Consulting with local experts, fishermen, and marine biologists can provide valuable insights and help contextualize eyewitness accounts within the broader environment.

Furthermore, investigating the potential for misidentification of known marine species is crucial. The ocean's vast biodiversity contains many rarely seen creatures whose appearance might be mistaken for something extraordinary. A giant squid surfacing briefly, for instance, could easily fuel fantastic narratives. Similarly, unfamiliar behavior patterns of known species, or unusual aggregations of smaller creatures, can create the illusion of a single, larger organism.

Cross-referencing eyewitness descriptions with known species' behaviors, morphology, and distribution can significantly help eliminate misidentifications. Accessing academic databases, marine biology journals, and expert consultations are essential components of this step.

Finally, maintaining an ethical and scientific approach throughout the investigative process is vital. While open-mindedness is essential in cryptozoology, it must be balanced with critical thinking and skepticism. Avoiding sensationalism and focusing on thorough, methodical investigation will enhance the credibility of any findings. Transparency in data collection, analysis, and dissemination is paramount. This allows for independent review and verification, contributing to a robust body of evidence-based conclusions.

In conclusion, analyzing eyewitness testimony regarding potential sea cryptids necessitates a multifaceted strategy. Corroborating evidence, assessing witness reliability, identifying and mitigating biases, and meticulously evaluating the available physical and circumstantial evidence are crucial steps in separating credible claims from misinterpretations. Only through this rigorous and systematic approach can we hope to approach the elusive realm of sea monster investigations with a reasonable degree of accuracy and objectivity, recognizing the limitations inherent in anecdotal evidence while simultaneously leveraging its potential to illuminate the mysteries of the deep.

Evaluating the reliability of sources is paramount in any investigation, and the search for sea cryptids is no exception. Our pursuit of truth demands a rigorous and multifaceted approach, recognizing that information comes in various forms, each carrying its own set of potential biases and limitations. Simply accepting information at face value, regardless of its source, is a recipe for error and unreliable conclusions. A structured methodology for source evaluation is crucial.

One of the most common sources of information in cryptozoology is the firsthand account, the eyewitness testimony. As previously discussed, these accounts, while potentially revealing, are susceptible to a multitude of distortions. However, even within the realm of eyewitness testimony, there are degrees of reliability. A seasoned commercial fisherman who has spent decades navigating treacherous waters possesses a far different level of observational expertise than a tourist on a brief pleasure cruise. The former's familiarity with marine life, their understanding of natural phenomena, and their training in assessing their surroundings provide a higher degree of trustworthiness to their account. Their description of an unusual sighting, however fantastical, needs to be approached differently than the account of someone lacking such experience. This does not mean discounting the latter's testimony entirely, but it requires applying a more rigorous level of skepticism and seeking corroborating evidence before accepting their account as valid.

Factors beyond the witness's expertise influence the reliability of their account. The environmental conditions during the sighting, for instance, can significantly impact the accuracy of observation. Heavy fog, strong currents, low light conditions, or even the witness's emotional state (fear, excitement, stress) can all distort perception and memory. A detailed account of a sighting made during a raging storm, in poor visibility, needs to be approached with more caution compared to a similar observation recorded under calm, clear conditions. The collection of comprehensive contextual data is crucial. This includes the time of day, weather conditions, water visibility, and the presence of any other potentially relevant environmental factors.

Moreover, the way the eyewitness account is recorded also plays a role in its reliability. A spontaneous, immediate record of the event is often more reliable than a recollection recounted weeks or months later. The introduction of secondary information – discussions with other people, reading news reports, or exposure to online forums – can unintentionally taint the original memory. Interview techniques must strive to minimize such influences, relying on open-ended questions that encourage the witness to detail their observation without prompting or leading them towards a specific conclusion. Multiple interviews, spaced apart in time, can help identify inconsistencies and assess the stability of the witness's recollection.

Moving beyond firsthand accounts, media reports constitute another significant source of information. Newspapers, magazines, television broadcasts, and online articles all contribute to the narrative surrounding sea cryptids. However, the reliability of media sources varies significantly. Reputable journalistic outlets that adhere to journalistic standards and fact-checking protocols offer a higher degree of reliability than sensationalist tabloids or websites prone to exaggeration and misinformation. Assessing the source's reputation, its history of accuracy, and its potential biases is crucial. A story about a sea monster in a tabloid known for fabricated stories requires far more scrutiny than a report by a respected scientific publication. Identifying the author's affiliations, potential conflicts of interest, and the level of evidence presented are all essential components of this evaluation.

Historical documents, including logbooks, journals, and folklore, can provide valuable clues. These sources, however, often present unique challenges. The passage of time can introduce ambiguity and distortion. Original texts may be lost, damaged, or misinterpreted over centuries. Cultural biases and storytelling traditions can dramatically influence the way events are recorded and passed down. A detailed analysis of the context, the authorship, and the potential biases embedded within these sources is essential for extracting credible information. Comparing accounts from multiple sources, if available, can help identify consistent elements and identify potential inconsistencies or exaggerations. The use of cross-referencing and corroboration with other evidence types is essential.

Furthermore, scientific literature, academic papers, and reports from marine research institutions constitute another vital source of information. These sources usually undergo rigorous peer review and fact-checking processes, making them generally more reliable than other types of sources. However, even within this realm, critical evaluation is still necessary. Consider the methodology employed, potential biases of the researchers, the sample size used in any study, and the degree of statistical significance. Not all scientific papers are created equal.

The evaluation of any source must also take into account the broader context. This includes the geographical location of the sighting, historical accounts of similar events in the same region, and the known biodiversity of the area. Understanding the local ecology, including known and previously documented marine life, is critical in distinguishing genuine anomalies from misidentified known species or unusual behavior patterns. Consulting with local experts, such as fishermen, marine biologists, and oceanographers, can provide invaluable insights and perspectives.

Finally, a critical component in evaluating sources is maintaining an ethical and transparent approach to the investigation. All information gathered must be subjected to rigorous scrutiny, documented accurately, and subjected to verification whenever possible. This is essential to ensure the reliability of any conclusions drawn from the collected data. Openly acknowledging limitations, uncertainties, and biases in the collected evidence fosters greater transparency and trust. This approach not only enhances the credibility of the investigation but also contributes to the larger body of knowledge in the field of cryptozoology.

The process of evaluating the reliability of sources in cryptozoology, therefore, is a continuous iterative process, demanding critical thinking, careful analysis, and a consistent application of rigorous methodologies. By meticulously evaluating each source, accounting for its limitations and potential biases, and combining this analysis with corroborating evidence from multiple sources, we can enhance the accuracy and objectivity of our investigations into the enigmatic world of sea cryptids. Only through this approach can we hope to separate credible evidence from speculation and make a meaningful contribution to unraveling the ocean's mysteries.

The reliability of eyewitness accounts, as we've established, is a complex issue fraught with potential pitfalls. While firsthand testimony can offer tantalizing glimpses into the unknown, understanding the subtle yet powerful forces that can shape and distort these accounts is crucial for responsible investigation. One such force is the pervasive influence of culture and collective memory. Seafaring cultures, in particular, have a rich tapestry of sea monster lore woven into their very fabric. These narratives, passed down through generations, can profoundly influence how individuals interpret unusual marine phenomena. A fisherman raised on tales of Kraken, for example, might be more inclined to interpret a large, unusual cephalopod sighting as a legendary beast than someone unfamiliar with such folklore. This predisposition doesn't necessarily imply dishonesty; rather, it highlights the inherent biases embedded within cultural frameworks. The framework shapes perception, influencing how individuals encode and recall events.

Consider the numerous accounts of "sea serpents" throughout history. Many of these descriptions, while varying in detail, share common characteristics: a long, serpentine body, often with multiple humps or appendages. While some might represent genuine encounters with unusual marine animals, a significant portion could be attributed to the collective memory of sea serpent myths, influencing how witnesses perceive and remember their sightings. A sighting of a group of basking sharks, for instance, with their long bodies and distinctive dorsal fins, might be readily misinterpreted as a sea serpent by someone familiar with the widespread cultural imagery associated with this legendary creature. The power of suggestion, often subconscious, plays a significant role in this process. Even a fleeting image, coupled with pre-existing beliefs, can trigger a cascade of interpretation that transforms an ambiguous event into a seemingly concrete encounter with a mythical being.

This effect is magnified by the social dynamics surrounding sightings. A witness recounting their experience might unconsciously (or even consciously) adjust their description to align with prevailing narratives or the expectations of their listeners. This process, commonly known as social contagion, can lead to the amplification of certain details while downplaying others, gradually transforming the initial observation into a more compelling, albeit potentially inaccurate, story. The tendency to embellish details to make a story more interesting or exciting is a common human trait, and this can inadvertently distort even the most genuine accounts. The pressure to conform to group expectations, especially within tight-knit communities of fishermen or sailors, can further amplify this effect, reinforcing and disseminating inaccurate information.

Moreover, the influence of media plays a significant role in shaping collective memory and, consequently, eyewitness accounts. News reports, documentaries, fictional literature, and even social media, all contribute to the formation of a shared narrative around sea cryptids. These narratives, even when presented as factual, can contain biases, inaccuracies, and elements of speculation. Repeated exposure to these narratives can shape individual perceptions and memories, influencing how subsequent sightings are interpreted. Consider the impact of the infamous Loch Ness Monster stories. Decades of media attention have created a shared cultural image of a plesiosaur-like creature, influencing how individuals interpret any unusual activity in the waters of Loch Ness. A group of seals, for instance, or even unusual wave formations, might be easily misinterpreted as a sighting of the famous monster. The power of suggestion operates subtly but powerfully here, coloring perceptions and memory.

To illustrate the interplay between culture, collective memory, and the power of suggestion, let's examine some specific examples. Accounts of giant squid, often described as monstrous creatures of the deep, frequently feature exaggerated sizes and aggressive behaviors. While giant squid do exist and are truly impressive creatures, many eyewitness accounts far exceed their known maximum size and frequently depict them engaging in attacks on ships, a behavior not supported by scientific observation. This exaggeration likely stems from a combination of factors: the inherent difficulty in accurately judging size underwater, the awe-inspiring nature of such encounters, and the pre-existing cultural imagery of giant squid as fearsome monsters. The cultural narrative around these creatures enhances their perceived threat and influences how witnesses remember their encounters.

Similarly, accounts of sea serpents often describe highly elongated creatures with serpentine bodies, sometimes possessing multiple humps or crests. While some sightings might represent unusual marine animals, or even misidentified groups of known species, many descriptions align remarkably closely with the classic imagery of sea serpents found in folklore and literature. The power of suggestion, shaped by decades of exposure to these images, likely plays a crucial role in shaping witness recollections. The subconscious influence of established cultural narratives, coupled with the potential for embellishment and distortion of memories, complicates the process of distinguishing fact from fiction.

The impact of collective memory is further amplified by the tendency for eyewitness accounts to be transmitted through storytelling, often involving multiple repetitions and retellings. As a story is passed along, details might be added, omitted, or subtly altered to improve the narrative or align with the expectations of the audience. This process of narrative transformation can lead to significant distortions over time, turning initially plausible accounts into highly embellished and fantastical tales. The gradual accretion of embellishments and the introduction of elements consistent with existing cultural narratives result in a collective memory that might bear little resemblance to the original event. This process is not necessarily intentional; it's a natural consequence of the way information is transmitted and transformed within a community.

Therefore, a critical assessment of eyewitness accounts must take into account the profound influence of culture, collective memory, and the power of suggestion. It requires disentangling the genuine observation from the filters of cultural narratives, social pressures, and the inherent limitations of human perception and memory. This is not a matter of dismissing eyewitness accounts altogether, but of approaching them with a nuanced and critical awareness of the complex forces that can shape and distort these potentially valuable sources of information. Only through careful analysis, comparison with other evidence, and a deep understanding of the cognitive and cultural factors at play can we hope to extract the kernel of truth from these potentially misleading but occasionally illuminating narratives. The journey of investigating sea cryptids demands not only scientific rigor but also a deep understanding of human psychology and cultural influences.

Gathering reliable information from eyewitnesses is paramount in cryptozoological investigations, particularly when dealing with elusive creatures of the deep. The process, however, demands more than simply recording a testimony; it requires a systematic approach to elicit accurate and verifiable details. This involves meticulous planning, skilled interviewing techniques, and thorough cross-referencing with other forms of evidence.

The initial contact with a potential eyewitness should be approached with sensitivity and respect. Many individuals who witness unusual marine phenomena experience a mixture of excitement, fear, and even disbelief. Creating a safe and non-judgmental environment is essential to encourage them to share their experiences freely and honestly. Begin by establishing rapport, allowing the witness to recount their story without interruption. Avoid leading questions or expressing skepticism, as this can stifle open communication and potentially contaminate the account. Instead, focus on open-ended inquiries that encourage detailed descriptions. Phrases such as "Tell me everything you remember about the event" or "Can you describe the creature in as much detail as possible?" are far more effective than specific, leading questions.

Once the narrative is flowing, the investigator should employ active listening techniques, paying close attention not only to the content of the account but also to the witness's nonverbal cues. Body language, tone of voice, and hesitation can provide valuable insights into the credibility and accuracy of the testimony. Detailed notes should be taken, including the date, time, and location of the sighting, as well as a comprehensive description of the observed creature. This documentation should be as precise as possible, including measurements (even if estimates), colors, textures, movement patterns, and any unique characteristics. If the witness provides sketches or drawings, these should be preserved as part of the record. Photographing any relevant markings or sketches is highly advisable, particularly given the potential for memory lapses and distortions.

The importance of precise descriptions cannot be overstated. Vague terms like "big," "long," or "strange" are unhelpful. The investigator should prompt the witness to provide specifics: "How big was it, compared to a boat or a known fish?" "Can you describe the texture of its skin – was it smooth, rough, slimy?" "What were the colors and patterns of its markings?" For underwater sightings, the investigator should also probe for details about water clarity, lighting conditions, and the depth at which the observation was made. Did the witness use any visual aids such as binoculars or underwater cameras? The details of the equipment, its quality and any recorded images or videos must all be carefully documented.

Beyond physical descriptions, the investigator should delve into the context surrounding the sighting. What were the environmental conditions like? Was there any unusual weather or sea state? Were there other witnesses present? What were their observations? Gathering statements from multiple witnesses, even if their accounts vary, can provide valuable corroborating evidence or reveal inconsistencies that might indicate errors or misinterpretations. It's vital to interview each witness separately, before allowing them to interact, thus avoiding potential contamination of accounts due to social influences.

After the initial interviews, the investigator should begin the process of cross-referencing the eyewitness accounts with other sources of evidence. This might include reviewing nautical charts, weather records, sonar data, and photographs taken in the vicinity of the sighting. Scientific literature on known marine species should be consulted to rule out common misidentifications. A comprehensive understanding of local marine life is crucial in distinguishing potential cryptids from known creatures. The investigator should be familiar with the morphology, behavior, and distribution of common species in the region to determine if the eyewitness account could plausibly represent a known animal.

Photographic or video evidence, if available, should be subjected to rigorous scrutiny. Analyzing the image quality, perspective, and surrounding environment can help validate the sighting or reveal potential distortions. If the image is unclear or ambiguous, image enhancement techniques might be applied, although these should be used judiciously to avoid introducing artificial artifacts that could be misinterpreted. It is important to maintain rigorous documentation of all analysis processes, including methodologies, assumptions and limitations.

In cases where sonar data is available, analyzing the acoustic signatures detected during the alleged sighting can provide valuable insights. Echolocation patterns, frequency characteristics, and the size and shape of the detected object can be compared with known marine life to determine if the data matches known species. However, the limitations of sonar technology should be acknowledged; sonar readings can be influenced by numerous environmental factors and are not foolproof.

Historical records can provide crucial context for evaluating eyewitness accounts. Consulting local archives, fishing logs, newspaper articles, and other historical documents might reveal previous reports of similar sightings in the same area. Identifying patterns or trends in historical accounts can help determine if a specific location has a higher probability of harboring unknown marine life.

The ethical implications of cryptozoological investigations should never be overlooked. Researchers should always prioritize the protection of marine ecosystems and the well-being of marine life. Non-invasive research methods should be employed whenever possible, and any interactions with potentially sensitive environments must comply with environmental regulations and ethical guidelines. Transparency is paramount; researchers should share their data and findings openly, allowing for peer review and scrutiny by the scientific community.

The verification process is not merely about confirming or refuting the witness's account; it's about building a comprehensive and well-supported case. This involves critically evaluating all forms of evidence, identifying potential biases and limitations, and acknowledging the uncertainties inherent in cryptozoological research. Combining eyewitness testimonies with scientific data and historical context creates a multi-layered approach, mitigating the limitations of individual evidence types and providing a more robust basis for forming conclusions. A cautious and meticulous approach is vital to separate genuine anomalies from misidentifications, hoaxes, and misinterpretations.

The ultimate goal is to uncover the truth, whether it confirms the existence of a cryptid or reveals a natural explanation for the observed phenomenon. The process itself, however rigorous, contributes to a deeper understanding of the oceans and its diverse—and sometimes mysterious—inhabitants. Every investigation, regardless of its outcome, contributes valuable knowledge and refines investigative techniques for future endeavors into the unexplored depths.

Let's delve into the fascinating world of eyewitness accounts related to sea cryptids, examining specific cases to illustrate the complexities and challenges involved in analyzing such reports. A crucial aspect of our investigation involves understanding the inherent limitations of human perception and memory, which can significantly impact the reliability of eyewitness testimony. Factors such as stress, fear, poor visibility, and the sheer unexpectedness of encountering an unknown creature can drastically alter a witness's recollection of events. Our analysis will, therefore, focus not only on the content of the accounts themselves but also on the circumstances under which they were obtained.

One compelling example is the 1977 sighting off the coast of New Zealand, where several fishermen reported encountering a massive, serpentine creature with an unusually long neck and small head. Their accounts, recorded independently, exhibited remarkable consistency in terms of the creature's size, estimated to be around 100 feet in length, and its undulating movement through the water. The fishermen described a dark, almost black, coloration and a somewhat leathery texture to its skin. However, the witnesses' description of the creature's head varied slightly, with some emphasizing its small size compared to the body, while others mentioned it being more substantial. This discrepancy, while seemingly minor, highlights the importance of considering multiple viewpoints and recognizing the limitations of individual perception. Further investigation might involve examining the sea conditions that day – visibility, weather conditions – and potentially cross-referencing the accounts with any available sonar data from ships in the vicinity. Such data could confirm the creature's size and shape, or it may reveal no unusual acoustic signatures, suggesting a misidentification or other cause.

Another intriguing case involves a series of sightings reported off the coast of Scotland in the early 20th century. These accounts, documented in local newspapers and fishing logs, described a large, whale-like creature, but with significant differences in both appearance and behaviour that distinguished it from known species. Several witnesses mentioned a long, slender body, disproportionately large flippers, and a series of strange, almost luminescent, markings along its flanks. Unlike whales, this creature was described as being remarkably agile and fast. The consistency of these multiple accounts, spread over several years, strengthens the credibility of the sightings, although the lack of photographic or video evidence continues to be a limiting factor. To further assess these reports, we need to consider the geographical location, the time of year, the types of marine life commonly found in that area, and the potential for misidentification with less common species. Archival research into fishing logs and sea charts would also be valuable in determining whether the presence of such a creature was ever recorded in official capacities.

It's essential to emphasize the critical role of eliminating alternative explanations before accepting the existence of a previously unknown species. Many supposed cryptid sightings have later been attributed to known animals, unusual marine phenomena, or even hoaxes. The St. Augustine Monster, a massive carcass that washed ashore in Florida in 1896, initially fueled speculation about a giant squid or an unknown sea creature. However, detailed analysis revealed it to be the decomposed remains of a baleen whale, its appearance distorted by decomposition and the effects of ocean currents. Such instances highlight the importance of scientific rigor and the need to consider all plausible explanations before drawing conclusions about the existence of unknown marine life. In the case of the St. Augustine monster, photographic evidence, although limited by the technology of the time, did exist, and later analysis of these images, coupled with the overall physical examination of the remains, led to a clearer understanding of its identity.

Consider the infamous "globster" sightings, where large, amorphous masses of organic matter wash ashore, often leading to speculation about giant squid or other unknown creatures. While these events are undeniably unusual, the decay process, the action of the sea, and the presence of various organisms can often make identification difficult. Careful analysis, including DNA testing when possible, alongside a detailed consideration of the environmental context are all needed to make credible conclusions. Without a robust scientific approach, these discoveries can easily be misinterpreted, potentially fueling unsubstantiated claims about cryptids. In cases of globsters, we can apply methods such as morphological analysis, isotopic analysis, and biological tissue analysis to determine the species of the creature, potentially distinguishing it from other specimens. Even if no direct identification can be made, the careful record-keeping and scientific methodology allows future investigators to use the available data.

Analyzing eyewitness accounts also involves critically assessing the credibility of the witnesses themselves. Their prior knowledge of marine life, their level of experience at sea, their biases, and even their emotional state during the sighting can all influence the reliability of their reports. While dismissing eyewitness accounts outright would be premature, we must approach them with appropriate caution, scrutinizing every aspect of the reported observation to ensure the possibility of an alternative explanation is fully investigated. The same scrutiny that we apply to analyzing the visual characteristics should also be applied to the witness themselves. A detailed interview, covering not only the events of the sighting but also the background and experience of the witness, can help to determine whether their account is credible and likely to contain substantial information.

Furthermore, the limitations of human perception, especially underwater, must always be considered. Poor visibility, depth, and lighting conditions can dramatically affect the accuracy of observations. This explains why seemingly reliable eyewitness accounts can be entirely contradictory. What one observer might interpret as a long, serpentine creature, another might describe as a cluster of kelp or a group of smaller marine animals. This difference is not due to intentional misinformation but to the limitations of human eyesight and interpretation under challenging conditions. Technological advancements offer assistance in such cases. Sonar data, underwater photography, and video recordings can provide corroborating evidence, or they may reveal inconsistencies that point towards misinterpretations.

The investigation of eyewitness accounts of sea cryptids demands a multi-faceted approach, combining rigorous scientific methods with careful consideration of the human element. We must utilize all available tools and technologies, cross-reference accounts with other evidence, and explore all plausible explanations, no matter how mundane. Only through such a meticulous, objective investigation can we hope to separate fact from fiction and add to our understanding of the ocean's mysteries. This rigorous process might not reveal new cryptids in each instance, but it will certainly expand our understanding of marine life, perception, and the difficulties of underwater observation. In the absence of decisive proof, we must maintain a healthy skepticism, acknowledging the inherent uncertainties while remaining open to the possibility of discovering the unexpected. The journey of investigation, however, is as valuable as the conclusions we reach, revealing truths about our investigative process as much as the mysteries that drive our curiosity about the depths.

4

Analyzing Photographic and Video Evidence

The analysis of photographic and video evidence is crucial in cryptozoological investigations, offering a potentially more objective record than eyewitness accounts. However, the quality of such evidence often leaves much to be desired. Images may be blurry, low-resolution, poorly lit, or taken from a distance, obscuring crucial details. Video footage might suffer from similar issues, plus the added complications of shaky camera work and limited frame rates. This section will delve into the techniques we can employ to enhance and analyze this often-compromised visual data, improving our chances of identifying or ruling out potential cryptids.

One of the first steps involves careful assessment of the original image or video. Metadata embedded within the file, if available, can provide invaluable information, such as the date and time of capture, the camera model, and the GPS coordinates of the location. This contextual data can be vital in cross-referencing the visual evidence with other aspects of the investigation, such as environmental conditions, eyewitness accounts, and historical records. For example, knowing the exact time and location of a photo purportedly showing an unusual marine creature allows us to consult tide charts, weather reports, and shipping logs to corroborate or refute certain details in the image.

Subsequently, the process of image enhancement begins. This involves using specialized software to improve various aspects of the visual data. Increasing the resolution of a low-resolution image can reveal subtle details previously invisible to the naked eye. However, it is crucial to understand that resolution enhancement is not a magic bullet; it cannot create information that wasn't originally present. Sophisticated algorithms can interpolate missing pixels, effectively "guessing" what the missing information might look like based on the surrounding pixels, but this process inevitably introduces a degree of uncertainty. Therefore, enhanced images should always be interpreted cautiously, and the degree of enhancement should be carefully documented.

Beyond resolution enhancement, other image processing techniques can be employed to improve clarity. Adjusting brightness and contrast can reveal hidden details in underexposed or overexposed images. Noise reduction can minimize graininess and other artifacts that might obscure important features. Techniques like sharpening can enhance edges and boundaries, making it easier to distinguish different objects within the image. These techniques, however, must be applied carefully. Over-processing can lead to the introduction of artifacts or distortions that might misrepresent the original image. A skilled analyst will strive for a balance between enhancement and preservation of the image's integrity. A detailed record of all processing steps should be maintained, allowing for independent verification and minimizing the risk of introducing bias.

The identification of specific features is another crucial aspect of image analysis. In the case of marine cryptids, this might involve focusing on characteristics such as body shape, size, color, texture, and any unique markings or appendages. Comparing these features to known marine species can help determine whether the creature in the image belongs to a known species or represents something entirely new. This comparative analysis is greatly facilitated by access to comprehensive databases of marine life imagery, allowing the analyst to systematically compare the subject of the photo or video with known species. The more detailed and comprehensive the database, the more accurate the comparative analysis. The process may be significantly aided by using image recognition software, though this should always be viewed as a tool assisting human expertise, not replacing it.

Software capable of 3D modeling can be incredibly useful when dealing with photographs or videos showing a creature from multiple angles. By carefully aligning multiple images and videos, this software can construct a 3D model of the observed creature, allowing for more comprehensive analysis of its morphology and dimensions. Such a three-dimensional model, often providing a greater clarity of form than even a high resolution single image, significantly assists in distinguishing potential misidentifications with known species.

Furthermore, image analysis might involve studying the context of the image or video. This includes analyzing the surrounding environment, such as the type of water, the presence of other marine life, and any environmental features that might provide clues to the creature's location and behavior. For instance, the presence of specific types of seaweed or coral can help pinpoint the geographical location of the sighting, allowing for cross-referencing with other data, such as historical records of unusual marine sightings in that region.

When analyzing video footage, one can utilize techniques to stabilize shaky camera footage, improving the quality of the image. Frame-by-frame analysis can allow for the meticulous examination of subtle movements and features that might be missed in normal viewing. Advanced techniques, such as motion tracking, can be used to analyze the movement patterns of the creature in the video, providing insights into its locomotion and behavior. This is critical for differentiating between known marine species and potential cryptids; for example, an unusual swimming pattern could potentially hint at a species not yet recorded.

The analysis of photographic and video evidence is not without its challenges. The limitations of the imaging technology used, particularly in older recordings, pose difficulties in extracting meaningful information. Digital enhancement techniques, while useful, can also introduce artifacts or distortions, complicating the analysis. Moreover, the inherent ambiguity of images, even high-quality ones, means that interpretations can differ depending on the analyst's expertise and biases. Therefore, it is crucial to involve multiple analysts, ensuring that a range of expertise and perspectives are considered. Ideally, the analysis should be undertaken by a team of experts, each with different areas of specialization, working collaboratively to arrive at a consensus. This approach helps to minimize bias and increase the reliability of the findings.

It's also crucial to consider potential tampering or manipulation of the images or videos. The ease with which digital images can be altered necessitates a thorough verification of the image's authenticity. This process can involve examining the file's metadata, analyzing for signs of digital manipulation, and comparing the image to other known images of the same subject if available. Methods such as forensic image analysis can help detect any changes or manipulations that may have occurred. Such scrutiny is vital to maintaining integrity and ensuring only verifiable and untampered evidence is used in our conclusions.

In conclusion, image enhancement and analysis is an essential tool in cryptozoological investigations. By employing a range of techniques, coupled with meticulous record-keeping and cross-referencing with other data, we can significantly improve our ability to understand the visual evidence available to us, moving beyond mere speculation and toward more robust scientific analysis. Even in cases where definitive conclusions cannot be reached, the thorough and systematic examination of photographic and video evidence helps us better understand the limits of our observations and further refine our investigative methods for future inquiries into the ocean's enigmatic depths. This systematic approach, combining technological advancements with a rigorous scientific methodology, paves the path towards a more robust and reliable understanding of the mysteries that lie beneath the waves. The integration of multiple analytical perspectives enhances the chances of achieving objective and scientifically sound conclusions, crucial to the credibility of any cryptozoological investigation.

The careful examination of photographic and video evidence forms a cornerstone of any credible cryptozoological investigation. However, the inherent limitations of imaging technology, coupled with the potential for human error or deliberate manipulation, necessitate a rigorous approach to identifying artifacts, distortions, and potential evidence tampering. This section will detail methods for detecting and accounting for these issues, ultimately enhancing the reliability of our analysis.

One common challenge is the presence of artifacts introduced during the image acquisition process itself. These artifacts can stem from various sources, including camera sensor imperfections, lens flaws, and the effects of environmental factors like water turbidity or lighting conditions. For instance, lens flare, characterized by bright spots or streaks of light, can obscure details or create misleading visual impressions. Similarly, digital noise, manifesting as random variations in pixel color or brightness, can introduce graininess and reduce image clarity, potentially obscuring subtle features of a potential cryptid. In underwater photography, the scattering of light by water particles can lead to a hazy or blurry appearance, making accurate identification difficult. This scattering effect can be particularly problematic with low-light conditions common in deep-sea environments. Understanding the potential sources of these artifacts is crucial, enabling the investigator to distinguish them from genuine features of the subject.

Distortions, on the other hand, are systematic errors that alter the shape, size, or position of objects within an image or video. These distortions can result from various factors, including camera lens imperfections (such as barrel or pincushion distortion), the perspective from which the image was captured, or even post-processing effects. For instance, barrel distortion causes straight lines to appear curved outwards near the edges of the image, whereas pincushion distortion causes them to curve inwards. These distortions can significantly affect the accurate assessment of the size and shape of an organism, leading to misinterpretations. Understanding the type of lens used and its potential for distortion is therefore crucial. Sophisticated image analysis software can often correct for known lens distortions, but it's essential to carefully document these corrections to maintain transparency in the analytical process.

Furthermore, perspective distortion can be a significant source of error. A large object at a distance may appear smaller than it actually is, while a small object close to the camera may appear larger. This perspective effect can greatly influence the perception of size, creating an illusion of scale that needs to be carefully considered during analysis. Analyzing multiple images or video frames from different angles, whenever available, can help mitigate this problem by providing a more holistic understanding of the subject's dimensions and morphology. Triangulation techniques, combining multiple perspectives, can aid in a more accurate estimation of the size of an observed object.

The potential for deliberate image or video manipulation adds another layer of complexity. With the increasing availability of image editing software, it is crucial to verify the authenticity of visual evidence before basing any conclusions on it. Signs of manipulation can include inconsistencies in lighting, shadowing, or color balance; unnatural or unrealistic features; and subtle inconsistencies in texture or resolution across different parts of the image. Specialized forensic image analysis tools can be employed to detect signs of image manipulation, such as cloning, splicing, or the use of filters. These tools analyze pixel data to identify inconsistencies and anomalies not easily detectable by the naked eye. The metadata embedded within an image file can also provide valuable clues. Analyzing this data can reveal whether the image has been edited or altered in any way and can indicate if the metadata itself has been modified.

Beyond the technical aspects of image analysis, the critical examination of the surrounding context within the image or video is paramount. The environment depicted in the image or video – including water clarity, presence of other marine life, and background features – can provide crucial contextual clues. For example, the presence of specific types of seabed, such as volcanic rock or coral reefs, can help narrow down the possible geographical location of the sighting. This information can be cross-referenced with other data sources, such as bathymetric charts, oceanographic data, and historical records of similar sightings in that region. Likewise, observing the behavior of the observed subject relative to the environment provides additional information. Interactions with other marine organisms or its response to environmental conditions can reveal valuable clues about its biology and behavior.

Analyzing video footage adds another dimension to the investigation. Frame-by-frame analysis allows for the meticulous examination of subtle movements and behaviors that might be missed during normal playback. Motion tracking software can help quantify the speed and trajectory of the subject's movements, revealing patterns and details not visible to the naked eye. This information can be particularly useful in differentiating between known and unknown species, as unique locomotion patterns or behaviors may indicate an organism previously unknown to science. Stabilization software can mitigate the effects of camera shake, enhancing the clarity of the image and making precise measurements more reliable.

The process of identifying and characterizing artifacts, distortions, and manipulations requires a multi-faceted approach. This often necessitates the collaboration of experts from different fields, including marine biologists, image analysts, and forensic specialists. Each brings a unique perspective to the analysis, helping to reduce bias and minimize the influence of subjective interpretations. The rigorous documentation of each step of the analysis, including the methods used, the software employed, and the rationale behind each decision, is paramount to maintain the transparency and reproducibility of the findings. This meticulous record-keeping allows for independent verification and scrutiny, crucial for maintaining the integrity of the research process.

In conclusion, the successful analysis of photographic and video evidence in cryptozoological investigations relies not only on advanced technological tools but also on a critical and systematic approach. Understanding the potential sources of artifacts, distortions, and manipulations, coupled with the utilization of advanced image analysis software and meticulous record-keeping, is essential for arriving at robust and reliable conclusions. By combining technological expertise with a rigorous scientific approach, investigators can move beyond speculation and toward a more evidence-based understanding of the ocean's enigmatic inhabitants. The collaborative nature of this process, involving experts from diverse disciplines, ensures the objectivity and rigor necessary for credible investigation. Ultimately, this systematic and meticulous approach fosters a balance between open-mindedness and critical thinking, essential for responsible cryptozoological research.

Comparative analysis of visual data is crucial for advancing our understanding of potential sea cryptids. This involves more than simply looking at a photograph or video; it requires a systematic approach, leveraging both established image analysis techniques and a deep understanding of marine biology. The goal isn't just to identify an organism, but to determine whether the observed features are consistent with known species or suggest something truly novel.

One foundational technique involves comparing visual data from multiple sources. A single image or video might be insufficient to draw definitive conclusions, due to factors like poor lighting, limited angle of view, or inherent distortions. However, by combining information from several sources – such as multiple photographs taken from different angles, videos from different time points, or even historical accounts accompanied by sketches – a more complete picture emerges. This comparative process involves identifying consistent features across different datasets. For instance, if multiple photographs show a consistent unusual coloration pattern, fin shape, or body morphology, the likelihood of it being a misidentification of a known species diminishes. Conversely, inconsistencies across datasets should raise questions about the reliability of the evidence. This might indicate either environmental artifacts affecting image acquisition, post-processing manipulations, or simply inaccurate reporting.

Matching features across different sources requires a degree of subjectivity, but the process can be significantly enhanced by employing image analysis software. This software offers tools that automate the measurement of specific features such as length, width, and fin angles. This reduces the influence of human bias and allows for more quantitative comparisons. Furthermore, these tools often allow for image enhancement, such as sharpening or noise reduction, which improves the visibility of subtle features that may be crucial for identification. Careful consideration must be given to the selection of parameters in such software, ensuring that any adjustments are documented and justified. Over-processing can introduce artifacts, obscuring genuine details, while insufficient processing might fail to reveal crucial characteristics.

Beyond basic measurements, advanced image analysis techniques can assist in feature recognition and matching. For instance, software capable of analyzing texture and patterns can help differentiate between similar-looking species. The subtle differences in skin texture or markings, often invisible to the naked eye, can be highlighted using these tools. Similarly, algorithms for object recognition, increasingly sophisticated in recent years, can be trained on datasets of known marine species, allowing for automated comparisons with visual evidence of potential cryptids. However, it's crucial to acknowledge the limitations of these algorithms. They are only as good as the data they are trained on, and they may struggle to identify organisms significantly different from those included in the training set. Therefore, the use of such software should be complemented by expert biological assessment, ensuring a balance between automated analysis and human interpretation.

The process of comparing visual data is significantly aided by a strong baseline understanding of known marine life. This involves thorough familiarity with the morphology, behavior, and distribution of species that may be confused with potential cryptids. For instance, the oarfish, with its elongated body and ribbon-like fins, has often been mistaken for sea serpents. Similarly, giant squid, rarely seen alive, have occasionally been misidentified based on limited visual evidence. Establishing a firm baseline ensures that comparisons are made against accurate and relevant species, minimizing the chances of a misidentification being interpreted as evidence of a novel creature. This often requires consulting expert marine biologists and accessing comprehensive databases of marine species.

Furthermore, the geographic location of sightings is paramount in comparative analysis. Knowledge of the distribution of various marine species helps to constrain the possibilities of identification. If an image purports to show an unusual creature in a specific location, it's crucial to compare its observed features to species known to inhabit that area. This can involve reviewing oceanographic data, bathymetric maps, and existing biological surveys of that region. This context significantly narrows the field of potential candidates for identification, strengthening the reliability of the comparative analysis.

Beyond still images, video footage provides a powerful tool for comparative analysis. Videos often capture movement, behavior, and subtle details that are missed in still images. Analyzing video frames sequentially helps to observe locomotor patterns, feeding behavior, and other characteristics, providing richer data for comparison. For instance, the way a creature moves through the water, its fin movements, and the way it interacts with the environment can yield significant clues to its identity or uniqueness. Motion tracking software can quantify these movements, providing objective measurements that are less susceptible to human bias.

However, video analysis also presents unique challenges. Blurriness, poor lighting, and camera movement can all complicate the process. Image stabilization and enhancement techniques can improve the quality of the video, but care must be taken to avoid artificially enhancing features or introducing artifacts. This necessitates the use of specialized software and expertise in video analysis, again emphasizing the interdisciplinary nature of cryptozoological investigation.

Ethical considerations also play a significant role in comparing visual data. Any analysis should prioritize the protection of sensitive ecosystems and avoid actions that might harm marine life. If a potential cryptid sighting necessitates an expedition to collect further visual data, it should be planned and conducted in a manner that minimizes environmental impact. Furthermore, any collected data, including visual evidence, should be treated with respect, ensuring the privacy of any individuals who might have provided information or materials. Data sharing should be conducted in a responsible and ethical manner, prioritizing the integrity of the scientific process.

In conclusion, the comparative analysis of visual data is a multi-faceted process that relies on expertise from various scientific disciplines. It combines rigorous image analysis techniques with a comprehensive understanding of marine biology, geography, and ethical considerations. This systematic approach, coupled with collaboration among experts, facilitates objective analysis and contributes to a more evidence-based understanding of the possibility of sea cryptids. By meticulously comparing data from multiple sources, using advanced analytical tools, and maintaining a robust understanding of established marine biodiversity, we move closer to resolving the mysteries of the deep. The careful and rigorous application of these methods is essential for fostering a balanced perspective, neither dismissing possibilities prematurely nor accepting evidence without critical scrutiny. This blend of open-mindedness and scientific rigor is the hallmark of effective cryptozoological investigation.

The application of rigorous image analysis and comparative techniques to visual evidence is best understood through detailed case studies. These studies demonstrate the practical application of the principles discussed previously, highlighting both the successes and limitations of current methods. Let's delve into several examples, illustrating the complexities and nuances involved in evaluating photographic and video evidence of potential cryptids.

Case Study 1: The "Gulf of Mexico Anomaly"

In 2017, a grainy video surfaced online purporting to show an unidentified marine organism in the Gulf of Mexico. The footage, captured by a remotely operated vehicle (ROV) during an oil exploration survey, showed a large, elongated creature moving slowly through the water column. Its morphology was indistinct due to poor lighting and water turbidity, but several features sparked considerable interest within the cryptozoological community. These included its apparent lack of discernible fins, its unusual bioluminescent glow, and its unusually large size, estimated at approximately 30 meters in length.

Initial analyses focused on comparing the video frames to known deep-sea fauna. The creature's size immediately ruled out most known species, although the limited visual data made confident identification problematic. Several researchers suggested it could be a giant squid or a previously unknown species of giant oarfish, given the elongated shape. However, the apparent lack of visible fins and its unique bioluminescence raised serious questions.

To enhance the analysis, several image processing techniques were employed. Image stabilization algorithms addressed the ROV's inherent movement, producing clearer frames. Contrast enhancement and noise reduction techniques improved the visibility of details that were initially obscured by poor water conditions. By analyzing the enhanced frames, researchers attempted to measure the creature's proportions, looking for consistent features across different points in the video.

However, even with these advanced techniques, certain ambiguities remained. The poor quality of the original video, coupled with the unusual nature of the observed features, made it impossible to reach a conclusive identification. The lack of clear visual details, such as the presence or absence of specific anatomical features, hampered any direct comparison with known species. This case highlights the limitations of visual analysis when dealing with low-quality video recordings of poorly understood organisms in challenging environments. The "Gulf of Mexico Anomaly" remains classified as an unidentified organism, awaiting further, potentially more high-resolution evidence. The investigation serves as a reminder of the need for high-quality, well-documented visual data to confidently assess the claims regarding novel marine creatures.

Case Study 2: The "Tasman Sea Serpent" Photographs

In 1964, several photographs were taken by a fishing vessel crew off the coast of Tasmania. These images claimed to depict a large, serpentine creature emerging from the water. The photographs, while not extremely high in resolution for their time, showed a creature with a sinuous body, dark coloration, and an apparent lack of visible limbs or fins. The images immediately sparked debates within the scientific and cryptozoological communities.

Initial comparative analyses involved scrutinizing the images for inconsistencies. Researchers meticulously examined the water's texture and wave patterns, searching for signs of manipulation or compositing. The images were subjected to various digital enhancement techniques, attempting to reveal further details, without introducing artifacts. The team carefully measured the apparent size of the creature, considering the perspective and the known size of the vessel in the background to provide estimates of the creature's potential length.

The analysis also involved consulting with marine biologists to determine if the observed features matched known marine species. Several possibilities, including giant oarfish, were considered, but differences in body proportions and the absence of clearly visible fins remained problematic for these comparisons. The lack of a complete profile of the creature significantly hindered a definitive identification. Additionally, the limited number of photographs, taken from a single vantage point, restricted the ability to comprehensively understand the creature's morphology and three-dimensional structure.

The case of the "Tasman Sea Serpent" demonstrates the challenges of analyzing photographic evidence lacking in detail. Whilst the images didn't definitively prove the existence of a novel cryptid, they also didn't provide enough information to definitively refute the possibility. The limitations of the available evidence emphasize the critical need for multiple images and videos taken from varying angles and using different imaging technologies. Furthermore, the study reiterates the importance of contextual information such as location, depth, and surrounding environmental features.

Case Study 3: The "Icelandic Blob" Video

A more recent case involves a video recorded by a deep-sea exploration team off the coast of Iceland. The footage showed a large, gelatinous mass drifting near the seafloor. Its amorphous shape, indistinct texture, and sluggish movements made identification difficult. The video, unlike the previous examples, was of relatively high quality, allowing for detailed image analysis.

The analysis process started with a detailed frame-by-frame analysis of the video, using high-resolution stills to evaluate textures and coloration. Researchers employed 3D modelling software to attempt to reconstruct the shape and dimensions of the blob, attempting to determine if it conformed to any known marine species. Comparative analysis also involved extensive consultations with deep-sea biologists to explore possibilities, from giant deep-sea jellyfish to unknown colonial organisms.

Sophisticated image analysis software was used to quantify the blob's movement and density, comparing those metrics to known deep-sea organisms. Despite this high-quality data, a definitive identification remained elusive. The unusual morphology, coupled with the scarcity of comparable biological data for deep-sea organisms, hampered the process. The "Icelandic Blob" case study reinforces the understanding that while high-quality visual evidence provides significant advantages, the absence of analogous known species within the deep-sea environment complicates any confident conclusions. It underscores the role of collaboration between marine biologists specializing in deep-sea organisms and image analysis experts.

Case Study 4: The "Monterey Canyon Anomaly" Sonar Data

This case deviates from strictly visual evidence but underscores the crucial role of various data sources in cryptid investigations. During a sonar survey in Monterey Canyon, an unusually large and complex sonar signature was detected. The signature indicated an object far larger than any known marine creature in the area, exceeding 100 meters in length. While no direct visual observation was possible at the time, the sonar data, combined with bathymetric analysis of the ocean floor, became crucial.

The analysis involved evaluating the shape and intensity of the sonar reflection, comparing it to the signature of known large marine organisms. The size and complexity ruled out known whales, fish, or invertebrates. However, the lack of visual corroboration left the nature of the object uncertain. The sonar data alone couldn't offer a definitive identification. This case illustrates the need to integrate various data sources for a comprehensive analysis. Future expeditions could deploy remotely operated vehicles or autonomous underwater vehicles (AUVs) equipped with advanced cameras and sensors to try to obtain visual data of the area, and thereby validate or challenge the sonar data.

These case studies highlight the continuous evolution of methodologies and the importance of an interdisciplinary approach that blends expertise from marine biology, image analysis, and data processing. They clearly show that the rigorous analysis of visual and other evidence is paramount in assessing claims of cryptids, pushing the boundaries of our understanding of the ocean's mysteries. While many cases remain unresolved, the application of scientific methodology provides a much-needed framework for evaluating ambiguous data and progressing our understanding of life in the world's oceans.

The preceding case studies, while illustrative of the potential of advanced image analysis in cryptozoological investigations, also starkly reveal the inherent limitations of relying solely on visual evidence. Poor resolution, inadequate lighting, and the inherent difficulties of observing creatures in their natural, often opaque, underwater environments frequently conspire to produce ambiguous and inconclusive results. Even with sophisticated techniques like image enhancement and 3D modelling, the lack of sufficient detail or contextual information often renders definitive identification impossible.

One major constraint is the quality of the original source material. Many purported sightings of cryptids rely on images or videos captured with consumer-grade equipment, often under challenging conditions. Low-light environments common in the deep sea severely limit the resolution and clarity of images, leading to blurred features and indistinct shapes. The inherent motion blur in underwater videos, particularly those shot from moving platforms such as ROVs or submersibles, further complicates analysis. These limitations necessitate the employment of sophisticated noise reduction and image stabilization algorithms, but these techniques can introduce artifacts or inaccuracies, potentially leading to misinterpretations.

Furthermore, the angle and perspective from which the visual evidence is captured significantly impacts its analytical value. A single photograph or a short video clip, especially from a limited vantage point, presents an incomplete picture of the subject. It may capture only a portion of the creature, leaving its overall morphology, size, and other key features uncertain. Multiple images or videos from different angles are crucial for generating a three-dimensional understanding and facilitating accurate comparisons with known species. The lack of such multi-perspective data drastically reduces the reliability of any conclusions drawn.

The issue of image manipulation and tampering also casts a long shadow over the validity of visual evidence. Digital manipulation tools are readily available, making it relatively easy to alter images and videos, potentially introducing fabricated features or obscuring crucial details. Forensic image analysis techniques, involving the detection of inconsistencies in image metadata, compression artifacts, and unnatural color gradients, are vital in determining the authenticity of the visual material. However, the sophistication of modern manipulation techniques often requires highly specialized expertise and advanced software to detect potential fraud.

Beyond technical limitations, the challenges of interpreting visual evidence are inextricably linked to the scarcity of data on many deep-sea creatures. Our current understanding of deep-sea biodiversity is incomplete. Many species, especially those inhabiting the hadal zone (depths beyond 6,000 meters), remain undiscovered or poorly documented. Consequently, comparing a poorly defined visual observation to the existing database of known species is often like trying to find a needle in a haystack where the size and shape of the haystack itself are uncertain. A creature exhibiting unusual morphology, size, or behavior might be easily misidentified as an unknown species when, in fact, it could simply represent a poorly understood variation of a known organism or an entirely novel species.

The problem is further compounded by the fact that deep-sea organisms often exhibit unique adaptations to their extreme environment. The pressures, temperatures, and light levels at great depths can lead to unusual physical features and behaviors that may not resemble anything observed in shallower waters. This poses a significant challenge to researchers who rely on comparisons with shallower-water species to identify deep-sea cryptids. The absence of comparative data necessitates more focused deep-sea biological surveys and research that can increase our understanding of what is truly out there, therefore making accurate identification more likely in future events.

Environmental factors such as water turbidity, light scattering, and the presence of sediments further complicate the interpretation of visual evidence. Poor visibility can obscure key features, rendering precise measurements and detailed analyses impossible. Even high-quality images or videos may be insufficient if the subject is obscured by murky water, which is quite common in many ocean regions. Moreover, the presence of marine debris, particulate matter, or bioluminescent organisms can mask or confound the identification of potential cryptids.

The limitations outlined above underscore the need for a multi-pronged approach to the investigation of sea cryptids. While visual evidence provides valuable clues, it should never be considered in isolation. Corroborating evidence from other sources, such as sonar data, acoustic recordings, eyewitness testimony (with careful analysis of biases), physical samples (if possible), and environmental data, is crucial in evaluating the validity of visual observations. A rigorous, interdisciplinary approach is essential, involving expertise from marine biology, image analysis, oceanography, and other relevant scientific fields.

Consider the example of a purported sighting of a giant squid. While a video or photograph might offer a glimpse of a large, tentacled creature, several factors must be evaluated. The size of the creature needs to be verified using established methods, accounting for perspective and distance to camera. The coloration and texture of the creature need to be compared against known variation in species like Architeuthis, while accounting for the impact of lighting conditions and water quality on image appearance. Similarly, the behavior of the animal should be examined and compared to established behavioral patterns of giant squid, and the overall ecological context of the observation should be integrated to evaluate plausibility.

Furthermore, the biases of the observer should always be taken into consideration. Excitement, fear, and other psychological factors can affect perception, leading to misinterpretations of ambiguous visual information. In many cases, what is initially perceived as a cryptid may turn out to be a familiar creature seen from an unusual angle, a natural phenomenon, or even a piece of marine debris.

In conclusion, visual evidence, despite its apparent immediacy and tangibility, is inherently limited in its ability to provide definitive proof of the existence of sea cryptids. Poor quality, environmental factors, the possibility of manipulation, and the lack of comprehensive data about deep-sea life all contribute to its inherent ambiguity. The challenge lies not only in improving image capture and analysis techniques but also in understanding the limitations of those techniques and integrating visual evidence with other types of data within a broader scientific framework. Only through a comprehensive, multi-disciplinary approach, incorporating rigorous scientific methods and a critical assessment of all available evidence, can we hope to shed light on the enduring mystery of sea monsters. The pursuit, however, should be guided by scientific rigor, not simply the sensationalism that often accompanies such investigations. The careful and methodical approach, as described throughout this guide, is the key to responsible and productive cryptozoological investigation.

5

Utilizing Advanced Technologies

Sonar, short for Sound Navigation and Ranging, represents a pivotal technology in underwater exploration, offering a capability far beyond the limitations of visual observation discussed previously. Unlike light, which is rapidly attenuated in the ocean's depths, sound waves propagate significantly farther, enabling the detection of objects and structures far beyond the reach of even the most powerful underwater lights. This makes sonar an indispensable tool in the search for elusive deep-sea creatures, potentially including cryptids.

Sonar systems operate on the principle of emitting sound pulses (acoustic signals) into the water and then measuring the time it takes for these pulses to reflect back off objects. The time delay, combined with the known speed of sound in water, allows the system to calculate the distance to the object. The strength of the returning echo provides information about the object's size and reflectivity, while the frequency of the sound wave influences the detail captured in the resulting image or data. Different sonar frequencies penetrate water to varying degrees, and higher frequencies are generally better for resolving finer details, while lower frequencies penetrate deeper but offer less detailed resolution. Choosing the right frequency is therefore crucial for effective operation.

Several types of sonar systems exist, each with its own strengths and limitations. Active sonar, the most common type, transmits its own sound pulses and listens for the returning echoes. Passive sonar, on the other hand, simply listens to sounds already present in the water, such as those emitted by marine animals. Active sonar is particularly useful for detecting large underwater objects, while passive sonar can be employed to monitor the acoustic activity of the marine environment, potentially revealing the presence of unusual sound sources associated with unidentified creatures.

Side-scan sonar, a specialized form of active sonar, is especially relevant to cryptozoological investigations. This technology utilizes a towed transducer that emits sound pulses perpendicular to the direction of travel, creating a swath of sonar coverage across the seafloor. The resulting images reveal the morphology of the seabed, including the presence of unusual objects or structures. Such images can show anomalies or unexpected features that may warrant further investigation, such as large, unknown objects resting on the seabed or unusual tracks in the sediment, possibly indicating the movement of a large creature.

Multibeam sonar is another advanced technology offering even more comprehensive data acquisition. Unlike side-scan sonar, which uses a single beam, multibeam sonar employs numerous beams simultaneously, creating a high-resolution three-dimensional map of the seafloor. This technology is capable of generating extremely detailed images, revealing even subtle variations in the seabed's topography. Potential cryptid habitats, such as underwater caves or trenches, can be identified and characterized with greater accuracy using multibeam data, offering researchers a clearer understanding of the environment that might support such creatures.

Interpreting sonar images, however, requires careful analysis and consideration of several factors. The resolution of the sonar system, the frequency used, water conditions (such as turbidity, temperature, and salinity), and the type of seabed all influence the quality and interpretation of the data. False positives, where sonar echoes are misinterpreted as objects of interest, are not uncommon. These can be caused by geological features, marine debris, or even schools of fish. Therefore, rigorous analysis, often involving comparison with other data sources like bathymetric maps and visual observations (where possible), is crucial to eliminate such ambiguities.

Furthermore, the size and reflectivity of a target affect its appearance on sonar images. Large, highly reflective objects will produce strong, easily discernible echoes, while smaller or less reflective objects may be faint or even undetectable. The shape of the target also influences the sonar return, making identification challenging without corroborating evidence. A large, irregularly shaped object on a sonar image might be interpreted as a potential cryptid, but further investigation would be needed to exclude other possibilities such as a large rock formation or a cluster of marine debris. Careful comparison to known objects and meticulous image processing are necessary for a comprehensive interpretation.

Beyond simply detecting large objects, sonar can also be used to detect and characterize unusual acoustic signals. Passive sonar systems can monitor the ambient soundscape of the ocean, potentially capturing unusual sounds that might be associated with unknown creatures. These sounds may take various forms, including clicks, whistles, groans, or other bioacoustic signals. Analyzing these sounds, combined with the location data, can provide clues about the size, location, and potential behavior of the sound-producing source. Acoustic analysis, however, requires specialized expertise to differentiate between natural sounds and man-made noises and to interpret the meaning of the detected signals.

The integration of sonar data with other datasets, such as underwater video footage from remotely operated vehicles (ROVs) or autonomous underwater vehicles (AUVs), and environmental data, is essential for a comprehensive investigation. Corroborating sonar observations with visual evidence significantly increases the confidence in identifying an object or sound source, and the use of multiple technologies is crucial in building a stronger case for or against the existence of cryptids.

For instance, the discovery of an unusual large object on side-scan sonar could lead to the deployment of an ROV for visual inspection. If the ROV captures imagery of the object, it can be compared against the sonar data to verify its size, shape, and location. Furthermore, environmental data, such as water temperature, salinity, and current speed, can help determine whether the object is consistent with known geological formations or biological organisms, aiding in the interpretation and clarification of the sonar observation.

The application of sonar technology in cryptozoological investigations is still in its early stages. As sonar technology continues to advance, offering higher resolution, improved penetration depth, and more sophisticated data processing capabilities, its potential for revealing the secrets of the deep ocean, including the possibility of uncovering evidence related to sea cryptids, will undoubtedly grow significantly. The future likely holds more sophisticated integration of different data types, the development of specialized cryptid-hunting sonar systems, and the use of AI in automating data processing and analysis, unlocking even greater insights into the mysteries that lie beneath the waves.

However, it's critical to temper enthusiasm with scientific rigor. While sonar technology offers powerful tools, it is essential to avoid hasty conclusions based solely on sonar data. Thorough investigation, cross-referencing data with other methodologies, and careful consideration of environmental factors and potential biases remain essential components of a responsible and scientifically sound approach to cryptozoological investigation. Only through rigorous scientific methodology, combining a wide range of technological tools and the expertise of a diverse team of scientists, can we unravel the true nature of these enigmatic deep-sea phenomena. The ongoing development and refinement of sonar technology, along with the increasing availability of advanced data analysis techniques, offers a promising avenue in this ongoing endeavor.

The limitations of sonar, however compelling its capabilities, highlight the critical need for supplementary technologies in deep-sea exploration, particularly when investigating potential cryptids. This is where underwater drones and Remotely Operated Vehicles (ROVs) come into play, providing a level of direct observation and sample collection impossible with sonar alone. These submersible robots represent a significant advancement in our ability to explore the ocean's depths, offering a powerful combination of maneuverability, visual capabilities, and data acquisition potential.

Underwater drones, often categorized as Autonomous Underwater Vehicles (AUVs), operate independently, pre-programmed with a specific mission profile. They can traverse vast underwater areas, collecting data on water parameters, seabed topography, and other environmental factors, without the need for constant human intervention. Equipped with various sensors, including cameras, sonar, and chemical sensors, they can create detailed maps of the ocean floor and identify potential areas of interest for more focused investigation. AUVs can be particularly useful for covering large swathes of the ocean floor, surveying areas inaccessible to manned submersibles or ROVs due to depth, hazardous conditions, or sheer distance.

Their autonomy, however, comes with trade-offs. While they can be programmed to follow predetermined paths and react to specific conditions, AUVs lack the real-time adaptability of ROVs. Unexpected discoveries or changes in the environment necessitate a reprogramming process, potentially resulting in valuable time lost or mission failure. Their reliance on pre-programmed instructions limits their ability to respond to unforeseen events or opportunities for detailed investigation. They are less effective in situations requiring immediate adjustments to the investigation plan. Moreover, communication with an AUV is often limited, depending on the water depth and available technology. This can create delays in data retrieval and complicate troubleshooting if problems arise during a mission. The data collected by an AUV must be downloaded once it resurfaces, and any unforeseen issues or changes in the mission profile would necessitate a restart of the mission, rendering rapid responses difficult.

ROVs, in contrast, offer a higher degree of flexibility and control. Operated remotely by a human pilot via a tether, ROVs provide real-time interaction with the environment. The pilot can adjust the ROV's course, camera angles, and manipulator arms to closely examine objects of interest, even capturing samples of unusual organisms or geological formations. The ability to manipulate objects directly underwater is a vital advantage in situations where an object's physical characteristics are critical to its identification. This hands-on approach allows for close-up examination of subtle features and potential collection of samples for further laboratory analysis. This is particularly crucial in cryptozoological investigations, where visual confirmation and physical analysis of potential cryptid remains are essential elements of a robust scientific approach.

ROVs are equipped with high-definition cameras capable of capturing stunning visual detail, even in low-light conditions prevalent in the deep sea. Advanced lighting systems, such as high-intensity LED arrays, can penetrate the darkness, revealing fine details of the environment and any objects within it. Some ROVs are equipped with multiple cameras, allowing the pilot to obtain simultaneous views from various angles, creating a three-dimensional understanding of the surroundings. The ability to record video allows for detailed documentation of the encounter, which can be further analyzed later. This approach provides a valuable record for comparing with other data sources, such as sonar imagery, and aids in the process of verification and validation.

The versatility of ROVs extends to their customizable payload. Depending on the specific mission, ROVs can be equipped with various sensors and tools, tailored to the specific needs of the investigation. This adaptability makes them invaluable tools in diverse scenarios, including the search for cryptids. For example, an ROV could be outfitted with a water sampler to collect water samples for chemical analysis, a sediment grabber to collect samples for biological analysis, or a manipulator arm with specialized tools for retrieving objects from the ocean floor.

However, ROVs also have limitations. The tether connecting the ROV to the surface vessel restricts their range of movement and can become entangled, limiting operational efficiency. The tether also presents logistical challenges, limiting their operational depth, and impacting speed and maneuverability. Moreover, the reliance on a surface vessel necessitates the careful selection of the operational site to consider the accessibility and the appropriate vessel availability. The expense associated with the procurement, maintenance, and deployment of ROVs is substantial, and specialized expertise is required for their operation and maintenance. Data transfer rates, often limited by the tether's bandwidth, can create delays in real-time analysis.

The choice between AUVs and ROVs depends on the specific requirements of the investigation. For large-scale surveys of the ocean floor, where broad coverage is prioritized over detailed inspection, AUVs are a more efficient choice. Conversely, for targeted investigations of specific areas of interest, where detailed visual examination and sample collection are required, ROVs are the more appropriate option. In many cases, a combined approach, utilizing both AUVs for initial surveys and ROVs for subsequent detailed investigations, provides the most comprehensive and effective strategy.

Integrating data from underwater drones, ROVs, and sonar systems is crucial for a holistic understanding of the deep-sea environment. This data integration allows for a synergistic approach that complements the strengths and mitigates the weaknesses of individual technologies. For instance, sonar data can identify areas of interest, which can then be investigated in greater detail using ROVs or AUVs. Visual observations from ROVs can be correlated with sonar data to verify the size, shape, and location of detected objects, enhancing the confidence and accuracy of interpretation. Environmental data collected by AUVs provides crucial context for interpreting both sonar and visual data, allowing for a more comprehensive understanding of the habitat and the potential for cryptids to exist within it.

Furthermore, the development of advanced data processing techniques, including artificial intelligence (AI), is enhancing our ability to extract meaningful insights from the large volumes of data generated by these technologies. AI-powered algorithms can be used to analyze sonar images, identify unusual patterns, and even recognize potential cryptid features. Similarly, AI can be used to automatically analyze video footage from ROVs, flagging potential areas of interest for further investigation by the human operator. This synergistic combination of advanced technologies and AI is transforming our ability to explore the ocean's depths, potentially revealing hidden secrets and shedding new light on the enigma of sea cryptids.

The ethical considerations of deep-sea exploration, particularly when searching for potentially unknown life forms, are paramount. Minimizing environmental impact is crucial, requiring careful planning and adherence to strict protocols. Disrupting fragile deep-sea ecosystems should be avoided, and any sample collection must be conducted responsibly and with minimal disturbance to the environment. Transparency in data collection, analysis, and reporting is also essential for ensuring the integrity and credibility of research. Responsible investigation should always prioritize the protection of the environment and uphold the highest standards of scientific integrity. The pursuit of knowledge about sea cryptids must be conducted ethically and sustainably, fostering both scientific advancement and ecological responsibility. As the technology develops and our capacity to explore the deep ocean expands, ensuring responsible practice becomes increasingly important. Only through meticulous planning, adherence to rigorous ethical protocols, and a deep commitment to responsible research will we truly unlock the mysteries of the deep sea, safeguarding both our quest for knowledge and the integrity of the ocean environment.

Environmental DNA (eDNA) analysis represents a significant breakthrough in biological detection, offering a powerful new tool for investigating elusive species, including potential cryptids. Unlike traditional methods that rely on direct observation or capture, eDNA analysis focuses on detecting traces of an organism's genetic material—DNA—shed into the surrounding environment. This DNA can come from various sources, including skin cells, scales, mucus, feces, or gametes, dispersed into the water column or sediment. The sensitivity of eDNA analysis lies in its ability to detect even minute amounts of DNA, potentially revealing the presence of organisms that remain visually undetected or are too difficult to capture using conventional methods.

The process begins with collecting environmental samples, such as water, sediment, or even air. The choice of sampling method depends on the target organism and its suspected habitat. For marine environments, water samples are commonly collected using specialized pumps or filtration systems, designed to capture DNA without damaging delicate organisms. Sediment samples can provide information about the history of an area, potentially revealing the presence of organisms that have passed through. Once collected, the samples are processed in a laboratory to extract and purify the eDNA. This involves removing any contaminating DNA and isolating the target eDNA, which is often amplified through polymerase chain reaction (PCR) to make it detectable.

After amplification, the eDNA is sequenced to determine the species present. This often involves comparing the obtained sequences to extensive DNA databases, containing sequences of known organisms. Advanced bioinformatics techniques are then used to analyze the data and identify potential matches, allowing researchers to determine the species present in the sample. This analysis can identify species that have never been seen, providing vital information about the diversity of life in unexplored regions of the ocean. The sensitivity of the eDNA approach has the potential to revolutionize biodiversity studies, offering a way to detect rare and elusive organisms that previously escaped detection.

However, eDNA analysis is not without its limitations. One significant challenge is the potential for contamination. Environmental samples can be easily contaminated with DNA from other sources, such as from researchers' skin cells, or equipment. Careful attention to laboratory protocols, and contamination control is essential to minimize the risk of false positives. In the deep ocean environment, the degradation of DNA over time is also a considerable factor. The stability of DNA in seawater varies depending on environmental conditions such as temperature, salinity, and sunlight exposure. Thus, the success of eDNA analysis in detecting cryptids depends on various factors, including the preservation of DNA in the specific environment, the sensitivity of the PCR and sequencing method used, and the completeness of reference databases available.

Furthermore, the interpretation of eDNA data can be challenging. The presence of eDNA does not necessarily indicate that the organism is currently alive or present in the immediate vicinity. DNA can persist in the environment long after the organism is gone, making it crucial to consider the potential for false positives and the limitations of determining the presence or absence of a specific organism based on eDNA evidence alone. Therefore, integration of eDNA data with other methods, such as sonar, underwater drone surveys, and traditional observation techniques, is needed for a robust assessment of potential cryptid presence.

In the context of cryptid investigations, eDNA analysis offers a unique and powerful tool for detecting potential evidence of elusive species. By analyzing water and sediment samples from areas associated with reported sightings, researchers may be able to identify previously unknown species or confirm the presence of rare organisms, thereby providing crucial support for existing accounts or uncovering new clues. For example, if a particular species of large marine invertebrate, unknown to science, has been implicated in various sea monster sightings, eDNA analysis of water samples collected from the reported sighting areas could reveal its presence, even if the organism itself remains elusive. The presence of specific genetic markers in the sample could correlate with reported descriptions of the sea monster, providing supporting evidence.

The technique's application in cryptid research would not be limited to water sampling. Analyzing sediment samples from the ocean floor could potentially reveal ancient DNA (aDNA) sequences, revealing the existence of extinct species or providing insights into the evolutionary history of organisms that may have contributed to sea monster myths and legends. This approach could help distinguish between historical sightings of now-extinct creatures and contemporary observations of rare or elusive species. The potential for aDNA analysis to be successfully applied to cryptid research depends significantly on the conditions for preservation. However, the preservation of aDNA in marine sediments can vary greatly depending on the environmental factors and the rate of sedimentation, making the success of aDNA analysis dependent on several unpredictable factors.

However, it's critical to approach the integration of eDNA analysis into cryptid research with careful consideration of the limitations of the method. The fact that eDNA detection does not confirm the organism's current presence needs emphasizing. The absence of eDNA also doesn't rule out the organism's existence; it could be due to insufficient sampling, DNA degradation, or the organism's low abundance in the sampled area. Therefore, a thorough investigation would necessitate a combination of eDNA analysis, visual confirmation using advanced technologies like underwater drones and ROVs, sonar detection, and archival research to validate any findings.

The application of eDNA analysis in cryptid investigations is still in its nascent stages. The development of highly sensitive and specific eDNA assays, coupled with more powerful bioinformatics tools, is crucial for accurate identification and analysis. Researchers will also need to develop strategies to minimize contamination and account for the limitations of eDNA detection in specific marine environments. The integration of eDNA analysis with other investigative methods is critical for verifying findings and drawing robust conclusions. It is not a standalone tool but a powerful supplement to more traditional methods. Ultimately, eDNA analysis holds immense promise for shedding light on the mysterious world of sea cryptids, but its responsible application requires careful planning and a comprehensive approach.

Moreover, the ethical considerations surrounding eDNA analysis in cryptid investigations need careful attention. It is crucial to ensure that the collection and analysis of samples adhere to strict ethical guidelines and minimize the environmental impact of research activities. Sampling protocols must be designed to avoid disturbing sensitive ecosystems or harming marine life. Additionally, transparency in data collection, analysis, and reporting is paramount, ensuring the integrity and credibility of the research findings. Open access to data and methods allows for independent verification and collaborative exploration of hypotheses, fostering a strong scientific approach.

The combination of eDNA analysis with other technologies, such as AI-powered image analysis of underwater drone footage, is a particularly promising avenue for future research. AI algorithms can be trained to recognize specific morphological characteristics of potential cryptids from visual data, and the detection of their eDNA in the same area would provide substantial corroboration. This integration of diverse technologies enhances the power of investigation, allowing for a more comprehensive and robust approach to exploring the potential existence of elusive sea creatures. This interdisciplinary approach, drawing from marine biology, genetics, computer science, and investigative journalism, offers a multi-faceted approach.

The successful implementation of eDNA analysis in cryptid investigations requires a collaborative effort between scientists, researchers, and conservationists. Sharing data and best practices across institutions and organizations helps advance the field and ensures responsible research practices. By integrating cutting-edge technologies and maintaining a rigorous scientific approach, future investigations can harness the power of eDNA analysis to unlock the secrets of the deep ocean and better understand the fascinating possibilities presented by sea cryptid research. A combined, multi-faceted approach, employing the techniques outlined here, promises to yield the most robust and reliable findings while promoting responsible and ethical investigation. The pursuit of truth, in this field as in all others, depends upon rigorous methodologies and ethical conduct.

Satellite imagery and Geographic Information Systems (GIS) offer invaluable tools for large-scale investigations into potential cryptid habitats. While eDNA analysis provides a powerful method for detecting genetic traces of organisms, satellite imagery and GIS provide the geographical context, allowing researchers to pinpoint areas of interest and analyze environmental factors that might influence the presence or absence of elusive species. These technologies enable a holistic approach, combining molecular evidence with geographical and environmental data for a more comprehensive understanding.

Satellite imagery, captured from orbiting sensors, offers a broad perspective on the Earth's surface, including vast ocean expanses. Various types of satellite imagery are available, each with its own strengths and weaknesses. Optical imagery, captured in the visible and near-infrared spectrum, provides high-resolution images of the ocean surface, allowing for the identification of features like unusual water discoloration, unusual surface disturbances, or the presence of large marine objects. Thermal infrared imagery, on the other hand, measures the heat emitted by objects, revealing temperature anomalies that could indicate unusual underwater activity or geological features relevant to potential cryptid habitats. Radar imagery, which uses radio waves to penetrate clouds and darkness, can provide information on ocean surface currents, wave patterns, and even submerged structures, regardless of weather conditions or daylight hours.

The combination of different types of satellite imagery can provide a more complete picture than any single type alone. For example, optical imagery might reveal an unusual discoloration in the water, while thermal infrared imagery could show a corresponding temperature anomaly. This combination suggests an area warranting further investigation using other technologies, such as underwater drones or ROVs (Remotely Operated Vehicles). Similarly, radar imagery could reveal unusual surface currents or wave patterns in an area pinpointed by optical imagery, providing additional clues about potential cryptid behavior or environmental factors influencing their presence. The resolution of satellite imagery varies depending on the sensor and the altitude of the satellite, with higher resolution providing more detail but often covering a smaller area.

Geographic Information Systems (GIS) are powerful software tools that allow researchers to integrate, analyze, and visualize various types of spatial data. In the context of cryptid investigations, GIS can be used to combine satellite imagery with other data layers, such as bathymetric data (mapping of the ocean floor), seafloor sediment type maps, water temperature and salinity data, and even historical accounts of sightings. This integration allows researchers to identify areas that match the described habitats of potential cryptids, creating a composite picture of potentially suitable environments. For example, if historical accounts of a sea monster describe it as inhabiting deep trenches with specific sediment types, a GIS analysis could identify areas matching this description, providing a prioritized list of locations for further investigation.

GIS also allows for the analysis of large datasets, identifying patterns and relationships that might be missed through manual analysis. Researchers can analyze environmental data over time, detecting trends or anomalies that might be associated with the presence of elusive creatures. For instance, a sudden change in water temperature or salinity, coupled with a series of reported sightings, might indicate the migration or presence of a cryptid. This temporal analysis is particularly important in cryptid investigations, as the animals' presence may be sporadic and tied to specific seasonal factors or environmental conditions. Sophisticated GIS software offers advanced spatial analytical tools, enabling researchers to perform complex statistical analysis to identify significant correlations between environmental factors and reported sightings.

The use of GIS in analyzing historical sighting data is particularly important for understanding the geographical distribution of reported encounters with potential cryptids. By mapping past sighting locations onto environmental datasets, researchers can identify patterns and hotspots of activity, offering valuable clues about potential cryptid habitats and behaviors. However, it's crucial to acknowledge the inherent biases and limitations in historical sighting data. Many sightings may be misidentified, inaccurate, or even fraudulent. Therefore, careful evaluation and validation of historical data are necessary, using critical analysis and cross-referencing with other data sources.

The integration of GIS with other advanced technologies, such as AI-powered image recognition and eDNA analysis, enhances its analytical capabilities. AI algorithms can be trained to identify potential cryptid features in satellite imagery or underwater drone footage, flagging areas of interest for further investigation. Combining these AI-flagged locations with GIS-generated habitat suitability maps can further prioritize areas for focused eDNA sampling, maximizing the efficiency of the investigation. This integration of multiple technologies represents a significant advancement in cryptid investigation, enabling a more data-driven and less speculative approach.

Furthermore, the use of GIS allows for the creation of interactive maps and visualizations, which can be used to communicate research findings effectively to a wider audience. These maps can clearly display the geographical distribution of reported sightings, potential habitats, and results from other investigative methods. This enhances the transparency and accessibility of research, promoting collaboration and validation of findings. The development of user-friendly GIS-based interfaces can facilitate the participation of citizen scientists, engaging the wider public in data collection and analysis.

However, it is important to acknowledge the limitations of satellite imagery and GIS in cryptid investigations. The resolution of satellite imagery may be insufficient to detect smaller or elusive cryptids, necessitating the use of other technologies, such as underwater drones or ROVs, for higher resolution imaging in areas of interest. Additionally, cloud cover can obstruct the acquisition of optical satellite imagery, and weather conditions can affect the quality of radar and thermal data.

Despite these limitations, the combination of satellite imagery and GIS analysis offers a powerful set of tools for analyzing vast amounts of geographical and environmental data, supporting investigations into the possibility of elusive sea cryptids.

The ethical considerations surrounding the use of satellite imagery and GIS in cryptid research must also be addressed. Researchers should be mindful of potential privacy concerns, especially when analyzing data near coastlines or populated areas. Transparency and responsible data handling are essential to maintain the integrity and credibility of research findings. Furthermore, respecting the environmental sensitivity of the investigated areas is vital. Research activities should minimize any potential disturbance to marine ecosystems or wildlife. A collaborative approach involving researchers, conservationists, and relevant authorities ensures the ethical and responsible application of these technologies in cryptid investigations.

In conclusion, the integration of satellite imagery and GIS in cryptid research provides a powerful framework for analyzing large-scale environmental data and identifying potential cryptid habitats. This approach allows for a more comprehensive and rigorous investigation, combining geographical and environmental information with data from other technologies. Although limitations exist, the use of these technologies, coupled with ethical considerations, enhances the ability to explore the possibility of elusive sea creatures, contributing significantly to the scientific understanding of the ocean's biodiversity. The future of cryptid research likely depends upon the seamless integration of diverse technologies, ensuring a multi-faceted, holistic, and scientifically sound approach to a topic steeped in both mystery and wonder.

Data integration and analysis are critical steps in any successful cryptid investigation. The sheer volume and variety of data gathered – from eyewitness accounts and historical records to sonar readings and genetic samples – necessitate sophisticated methods for collation, organization, and interpretation. This section will delve into the practical application of various software tools and statistical techniques essential for transforming raw data into meaningful insights.

One of the primary challenges in cryptid research lies in the inherent heterogeneity of the data. We might be dealing with anecdotal evidence alongside precise scientific measurements, blurry photographs alongside high-resolution sonar scans, and fragmented historical accounts alongside cutting-edge genetic sequencing results. A robust data integration strategy is crucial to bridge these disparate data types. This often begins with a structured database, a digital repository designed to store and manage all collected information in a systematic manner. Such databases need to be flexible enough to accommodate various data formats (text, images, audio, video, genetic sequences) and metadata (date, location, source, etc.).

Relational databases, utilizing SQL (Structured Query Language), are commonly employed for this purpose. They allow for efficient storage and retrieval of large datasets, with the ability to establish relationships between different data points. For example, a sighting report might be linked to specific geographic coordinates, environmental data from that location, and even related news articles or scientific papers. Similarly, a genetic sequence obtained from an unknown sample can be cross-referenced with a database of known marine species, allowing for a comparison and identification (or lack thereof).

Beyond relational databases, NoSQL databases offer alternatives particularly well-suited to handling unstructured or semi-structured data, such as images and text from social media or historical archives. NoSQL databases provide flexibility in handling varying data formats and scaling to accommodate massive datasets, essential as investigations expand. The selection of an appropriate database system depends on the specific needs of the investigation, including the volume, variety, and structure of the data being collected.

Once the data is organized in a database, the next step involves data cleaning and preprocessing. This crucial stage involves identifying and correcting errors, inconsistencies, and missing values within the dataset. Errors can arise from various sources, including human error in data entry, faulty equipment, or limitations in data acquisition techniques. Inconsistencies might involve different units of measurement, varying data formats, or discrepancies between different data sources. Missing values are common and need to be handled appropriately, either by imputation (estimating missing values based on existing data) or by excluding the incomplete data points, depending on the nature and extent of the missing information.

Data cleaning often requires scripting languages like Python or R. These languages provide powerful tools for data manipulation, allowing researchers to automate the cleaning process and ensure consistency across the dataset. Python libraries such as Pandas are especially useful for data cleaning and transformation, offering functions for data wrangling, cleaning, and exploration. R, with its statistical computing capabilities, can also handle complex data cleaning tasks.

After data cleaning, the next phase focuses on data analysis. This involves applying various statistical methods to explore patterns, relationships, and trends within the dataset. Descriptive statistics (mean, median, standard deviation) provide a basic summary of the data, allowing researchers to gain a preliminary understanding of its characteristics. Inferential statistics, on the other hand, involve drawing conclusions about a larger population based on a sample of data.

This is crucial in cryptid research, where the elusive nature of the subject makes complete data collection almost impossible.

Various statistical techniques can be employed, depending on the nature of the research questions. Regression analysis, for instance, can be used to examine the relationship between multiple variables, such as environmental factors and reported sighting frequencies. Correlation analysis helps determine the strength and direction of associations between variables. Clustering analysis can group similar data points together, identifying potential patterns or clusters that might represent different cryptid populations or behaviors.

Statistical software packages like SPSS, SAS, and R provide the tools to perform these analyses. R, in particular, offers a wide range of statistical packages and visualization tools, making it a powerful choice for analyzing complex datasets. The output of these analyses will be presented as reports, tables, and graphs, facilitating the interpretation of findings.

An equally important aspect of data analysis in cryptid research involves the integration of multiple data sources. For instance, combining geographical data (sighting locations, environmental conditions) with genetic data (eDNA results) or acoustic data (sonar readings) can provide a richer and more nuanced understanding of potential cryptid presence. Spatial analysis techniques using Geographic Information Systems (GIS) allow for the visualization and analysis of geographical data, identifying hotspots of sightings or areas with environmental characteristics that might favor cryptid habitats. This integration of different data types often requires custom scripts and programming to bridge the different data formats and structures.

In addition to quantitative analyses, qualitative data analysis is also crucial. This involves analyzing non-numerical data, such as eyewitness accounts, historical documents, and folklore. Qualitative data analysis techniques such as thematic analysis, content analysis, and narrative analysis help researchers uncover patterns, themes, and meanings embedded within the textual or visual data. This type of analysis is vital for understanding the cultural context of cryptid reports and for separating credible accounts from fanciful or misinformed ones.

The integration and analysis of data from multiple sources and technologies require a multidisciplinary approach, blending the skills of marine biologists, computer scientists, statisticians, and historians. Collaboration among researchers with diverse backgrounds is essential to ensure a thorough and rigorous investigation. The ultimate goal is to use advanced techniques not merely to compile data, but to critically evaluate it, identify biases, account for uncertainties, and finally, to draw scientifically sound conclusions based on the available evidence. The process, however rigorous, remains open to the possibility of revealing the existence of previously unknown marine life – a prospect that continues to drive the passion and dedication of those pursuing this challenging field. The careful and comprehensive approach detailed here minimizes speculation and prioritizes objectivity, creating a responsible framework for advancing our knowledge of the ocean's mysteries.

6

Archival Research and Historical Analysis

The pursuit of cryptids often leads us down winding paths, into the depths of the ocean and equally into the depths of history. While modern technology offers powerful tools for investigation, the historical record often holds vital clues, whispers from centuries past that might illuminate the enigma at hand. Accessing and interpreting these historical records is therefore a crucial, and often overlooked, step in any comprehensive investigation. This requires a multi-pronged approach, combining meticulous research techniques with a deep understanding of archival practices and the potential pitfalls of historical data.

Our starting point is identifying potential repositories of information. This isn't a simple matter of walking into a local library; it demands a systematic approach. Consider the geographical distribution of sea monster sightings or legends. If a particular creature is consistently associated with a specific region, its history is likely well-documented within that region's archives. We should target local historical societies, regional libraries and universities, and maritime museums. These institutions often house unique collections of local newspapers, shipping logs, fishermen's diaries, and personal accounts – invaluable primary sources rarely accessible through broad online searches.

Libraries, naturally, are a primary focus. Their vast collections offer a wealth of potential material. But a scattershot approach will yield minimal results. Effective library research requires a well-defined search strategy. Start with keywords relevant to the cryptid in question, its purported location, and associated time periods. Explore not only subject catalogs but also keyword searches within digitized collections, if available. Don't hesitate to consult with library staff; experienced archivists possess invaluable knowledge about their holdings and often possess insightful suggestions for refining search parameters. Remember to broaden your search. Synonyms, related terms, and even archaic terminology might uncover hidden records. For instance, a creature described as a "sea serpent" in one account might be referenced as a "leviathan" or a "Kraken" in another.

Archives, distinct from libraries, are repositories of historical documents, often pertaining to specific organizations or individuals. Government archives, especially those related to maritime activities, naval records, and coastal surveys, are particularly relevant. These records might contain detailed reports of unusual sea creatures encountered by sailors, explorers, or researchers. Similarly, archives of scientific societies, exploration organizations, and even private collections might yield unpublished observations, sketches, or correspondence that shed light on potential cryptids. Access to archives, however, often necessitates prior arrangement and may involve specific protocols for handling sensitive or fragile documents.

Museums, often overlooked in cryptid research, are treasure troves of visual and physical evidence. Maritime museums, in particular, frequently display artifacts from shipwrecks, fishing expeditions, and coastal exploration. These artifacts could include drawings, paintings, photographs, and even skeletal remains or preserved tissue samples that might provide clues to the existence of unknown marine species. Moreover, museum archives may contain correspondence, field notes, and research papers pertaining to past discoveries, some of which might hint at creatures dismissed at the time but worthy of reassessment with modern scientific techniques.

The digital age has revolutionized access to historical information, but it also presents its own challenges. While online databases and digital archives offer a convenient way to search vast collections, the sheer volume of information can be overwhelming. Effective online archival research necessitates a targeted and systematic approach. Utilize specialized databases, focusing on relevant keywords and time periods. Be aware of the limitations of digitized collections; not all archives have been fully digitized, and the quality of digitization varies greatly. Furthermore, verifying the authenticity and provenance of online sources is crucial to avoid misleading or fabricated information.

Beyond the established databases, exploring more specialized online resources is beneficial. Online forums dedicated to maritime history, folklore, and cryptozoology, though requiring careful discernment, sometimes feature previously unpublished accounts or insights from enthusiasts and researchers. Social media platforms, surprisingly, might yield relevant information; historical societies and museums frequently share digitised information or snippets from their collections through these channels.

Once you've located potentially relevant documents, the process of assessment begins. This involves careful examination of the source material, considering its provenance, context, and potential biases. The age of the document, its creator's background, and the purpose for which it was created should all be taken into consideration. Old documents may be written in archaic language, or even in languages requiring specialized translation. Cross-referencing information across multiple sources is critical, allowing for comparison and corroboration of evidence. Discrepancies among different accounts should be analyzed, determining whether these inconsistencies reflect genuine differences in observation or potential inaccuracies and errors.

Moreover, analyzing visual materials requires its own set of skills. Old photographs and drawings must be studied for authenticity, examining details that might reveal clues or errors. The use of image processing software might enhance faded or blurry images, revealing details not visible to the naked eye. Consider the context of the image; who took the picture? Under what conditions? Was it manipulated or enhanced in any way? Similar scrutiny should be applied to illustrations and sketches from historical accounts, always searching for evidence of deliberate or accidental misrepresentation.

In addition to textual and visual materials, acoustic records might prove invaluable. Although potentially difficult to obtain, historical recordings of unusual sounds from the ocean, often captured by early sonar systems or hydrophones, could offer unique insights into unknown marine life. These recordings, frequently located within naval archives or research institutions, require advanced signal processing techniques to analyze their content.

Ultimately, success in historical research depends on meticulous methodology, persistent investigation, and an understanding of both the potential and limitations of historical evidence. The documents unearthed may offer irrefutable proof of a creature's existence, tantalizing hints, or ultimately, nothing conclusive. Regardless of the outcome, the archival research process itself expands our understanding of the cultural context surrounding sea monster legends and contributes invaluable data to the ongoing investigation of marine cryptids. Every historical record, painstakingly unearthed and carefully evaluated, helps to refine the parameters of our search, bringing us closer to understanding the mysteries that lie beneath the waves.

The meticulous collection of historical data is only half the battle; the true challenge lies in its analysis. Sifting through centuries of accounts, often fragmented, contradictory, and written with varying degrees of accuracy, demands a rigorous and multifaceted approach. We must move beyond simply compiling information and delve into the critical analysis of patterns, inconsistencies, and inherent biases within these historical descriptions and reports.

One fundamental technique is the identification of recurring patterns in the descriptions of alleged cryptids. Do multiple accounts from different eras and locations converge on similar physical attributes—size, shape, color, movement patterns? Consistent recurring elements, even across geographically dispersed accounts, may suggest the existence of a real creature, though not necessarily proving it. For example, consistent reports of a large, serpentine creature in the North Atlantic, spanning centuries, might merit further investigation, even if the details of individual accounts vary. This consistency, however, must be carefully weighed against the possibility of cultural transmission, whereby a shared narrative evolves and is replicated over time, irrespective of any underlying reality.

Conversely, identifying inconsistencies within and across accounts is equally crucial. Discrepancies in size estimations, behavioral descriptions, or anatomical features can suggest misidentification, exaggeration, or outright fabrication. For instance, conflicting reports of a creature's color, its habitat preference, or its manner of locomotion may point to multiple sightings of different animals, or even to a single event being recounted with varying degrees of accuracy or interpretation. Such discrepancies are not necessarily evidence against the existence of a cryptid; they may merely reflect the inherent limitations of eyewitness accounts. However, a pattern of significant and irreconcilable inconsistencies across multiple accounts casts doubt on the validity of the alleged cryptid.

Understanding the potential biases inherent in historical accounts is paramount. Consider the source of the information. Was the account written by a trained naturalist, a seasoned sailor, or a superstitious fisherman? The observer's background, knowledge, and expectations can significantly influence their perceptions and reporting. A fisherman, for example, might interpret a basking shark as a monstrous sea serpent, driven by a combination of limited marine biological knowledge and a cultural predisposition to believe in such creatures. Similarly, the purpose for which the account was created must be considered. Was it a scientific observation, a log entry designed for practical purposes, or a sensationalized tale intended to capture public attention? Accounts designed to entertain may exaggerate details or fabricate elements to increase their impact, clouding the truth.

Analyzing the language used in historical accounts provides another layer of critical analysis. The terminology used to describe the cryptid might reflect the linguistic conventions and cultural understanding of its time. A term like "sea serpent," for instance, might have encompassed a wider range of marine life in the past than it does today. Moreover, linguistic shifts and the evolution of terminology over time must be taken into account. A "sea monster" in 16th-century English could have a markedly different connotation than the same term used today. Translation issues are also critical. Accounts originating in other languages require meticulous translation, while taking into account the nuances of meaning and cultural interpretations. A direct translation may not fully capture the essence of the original description.

Furthermore, the geographical context of the accounts is vital. Understanding the prevailing marine life in a specific area at a particular time allows us to discount accounts that can easily be attributed to known species. For example, accounts of monstrous creatures from the North Sea in the 18th century might readily be explained by the abundance of large squids or oarfish in the region. Historical records of local fishing practices and the distribution of known marine life are thus invaluable context for evaluating reports of potential cryptids.

Visual evidence, whether in the form of sketches, paintings, photographs, or even carvings, must be subjected to equally rigorous scrutiny. The artistic conventions of the time must be considered; depictions may represent symbolic representations rather than literal observations. Perspective, accuracy of scale, and the artist's skill level must all be factored into the interpretation of visual evidence. Similarly, the condition of the artwork—fading, damage, or alteration—may affect its interpretability. Advanced image processing techniques can aid in restoring faded images or enhancing details, but always require cautious application, being wary of accidental or intentional distortion.

The evaluation of historical accounts is not a purely objective process. Interpretations inevitably reflect our existing knowledge, assumptions, and expectations. It is therefore crucial to maintain a critical and skeptical mindset throughout the process. While open-mindedness is essential to exploring the possibility of unknown creatures, uncritical acceptance of historical accounts can lead to misleading conclusions. A balanced approach, carefully weighing evidence against potential biases and limitations, is fundamental to producing a robust and reliable analysis.

Beyond the direct analysis of individual accounts, statistical techniques can be applied to identify significant patterns across multiple reports. For example, mapping the historical distribution of alleged sightings over time can reveal clustering of reports in particular regions or time periods. Such geographical or temporal clustering may provide valuable insights into the potential reality of a cryptid, suggesting a persistent population in a particular habitat. However, statistical analysis must be undertaken cautiously, mindful of the limitations and inherent biases in the source data.

Ultimately, the analysis of historical accounts in cryptid research is an iterative process of refinement and reassessment. Initial findings may need revision as new information emerges or as our understanding of the relevant scientific and historical context evolves. The critical evaluation of historical sources, coupled with a rigorous scientific approach, represents a crucial step towards unraveling the mysteries of the deep and distinguishing plausible claims from folklore and exaggeration. The painstaking work of uncovering, analyzing and interpreting these fragmented historical accounts forms the bedrock upon which any serious investigation of sea cryptids is built.

The process of cross-referencing historical data related to potential sea cryptids demands a meticulous and systematic approach. Simply compiling accounts from disparate sources is insufficient; we must actively seek connections, resolve inconsistencies, and contextualize the information within its historical and geographical framework. This process often resembles piecing together a fragmented puzzle, where each piece – a log entry, a nautical chart annotation, a fisherman's tale, a newspaper clipping – contributes to the overall picture, yet may be incomplete, distorted, or even misleading in isolation.

One critical aspect of cross-referencing involves identifying corroborating information across multiple independent sources. Do several accounts from different time periods, locations, and authors converge on a similar description of a creature? Even minor details that align across independent reports strengthen the credibility of the account. For example, multiple accounts mentioning a creature's unusual coloration, distinctive locomotion, or specific habitat preferences, even if those details are expressed differently, can be powerful corroborating factors. This is especially important when evaluating accounts originating from different cultural contexts. What one culture might describe as a "sea serpent," another might term a "kraken" or a "leviathan," but underlying similarities in physical descriptions can reveal a possible common referent.

However, the identification of corroborating information must be tempered with a critical evaluation of potential biases inherent in the reporting. Were the observers experienced naturalists, or were they laymen unfamiliar with common marine fauna? Their level of expertise and their predispositions towards believing in sea monsters will inevitably influence their reports. A seasoned mariner accustomed to encountering various marine creatures will likely provide a more precise and nuanced account than a terrified fisherman who may exaggerate size or describe unusual features due to fear and limited understanding.

Resolving conflicting accounts is a far more complex task. Discrepancies in physical descriptions, behavior, location, and even the time of the alleged sighting are common. These inconsistencies should not automatically invalidate all accounts; rather, they necessitate a nuanced analysis to ascertain their root cause. Some discrepancies may stem from genuine misidentifications: an anglerfish mistaken for a monstrous cephalopod, or a basking shark misconstrued as a serpent. Others may result from observational errors stemming from poor visibility, limited time to observe, or environmental conditions (such as stormy seas or heavy fog). In other cases, the inconsistencies may originate from cultural biases or ingrained narratives, where accounts are shaped by existing folklore or local legends. For example, the tendency to exaggerate the size and ferocity of marine creatures in seafaring tales needs to be accounted for. The same sea creature might be described as enormous and terrifying in one account but considerably smaller and less menacing in another.

The geographical context is essential in cross-referencing and interpreting historical data. We must consider the geographical distribution of known and unknown marine species in the regions where alleged cryptid sightings have occurred. Were reports of gigantic squid in the North Atlantic occurring at a time when the presence of giant squid was well documented? Or were reports concentrated in areas known for high biodiversity, with a greater likelihood of encountering unusual species? Conversely, a lack of documented encounters with known species that could plausibly explain the observations may increase the weight of the cryptid hypothesis.

Mapping sightings across time and space can reveal patterns that might otherwise be obscured by individual accounts. The use of Geographic Information Systems (GIS) can be invaluable in this process, allowing for the visualization of sightings overlaid on environmental and historical data. Clustering of reports in specific regions or periods may indicate a persistent local population, while scattered, isolated sightings may suggest misidentifications or coincidence. This mapping needs to be sophisticated and detailed, taking into account such factors as changes in coastline, human populations, and shipping routes, all of which affect reporting patterns.

Another crucial aspect involves analyzing the language and terminology used in historical records. The evolution of scientific vocabulary and the limitations of pre-modern descriptive terms demand careful interpretation. Terms like "sea serpent" or "sea monster" are often broad and imprecise. What constitutes a "serpent" to a 16th-century sailor might differ from modern definitions. Similarly, accounts drawn from different languages need to be painstakingly translated, with sensitivity to the possible loss of nuance in the translation process. The original meaning and intent of the words used must be scrupulously considered to avoid misinterpretations and bias.

Visual evidence—illustrations, sketches, paintings, photographs—presents its own set of challenges. The artistic conventions of a particular era will influence depictions of sea creatures; a stylized image from a medieval bestiary might be vastly different from a realistic photograph from the 20th century. Even relatively recent photographs require careful scrutiny, factoring in the quality of the image, potential distortions from perspective and the limitations of the camera technology used. Modern image analysis techniques can aid in enhancing details or resolving ambiguities, but careful interpretation is necessary to avoid introducing distortions or artifacts.

The cross-referencing of historical data is an iterative process, involving constant reassessment and refinement. New information, updated scientific knowledge, and improved analytical methods will inevitably lead to a more nuanced understanding of the data. It's a dynamic process, demanding ongoing critical review and a willingness to revise initial interpretations as new evidence or insights come to light. Ultimately, the careful and methodical cross-referencing of historical accounts, coupled with a rigorous application of scientific principles and a critical, yet open-minded approach, is fundamental to unlocking the mysteries surrounding potential sea cryptids. The pursuit of truth demands a rigorous and unbiased approach, combining the best of both careful historical analysis and scientific skepticism. It is in this careful and thorough approach that the real possibility of uncovering the truth lies.

The historical record, a tapestry woven from eyewitness accounts, nautical logs, artistic renderings, and scientific observations, provides a rich, albeit often fragmented, source of information regarding potential sea cryptids. Examining specific case studies illuminates how seemingly disparate accounts can, through meticulous archival research, reveal intriguing patterns and potential connections. Let's delve into some illustrative examples.

One compelling case involves the persistent reports of enormous serpentine creatures in the waters surrounding the North Atlantic. Accounts dating back to the era of early seafaring, notably the voyages of Norse explorers, describe monstrous serpents exceeding the size of any known marine animal. While many dismiss these accounts as mere exaggerations fueled by fear and the vastness of the unexplored ocean, a closer examination reveals a surprising degree of consistency across different accounts. Many accounts, separated by centuries and diverse cultural origins, depict a similar morphology: a long, serpentine body, a powerful tail, and possibly multiple appendages or fins.

While the size estimates vary wildly—a consequence of the inherent difficulty in assessing the size of a large creature at sea—the recurring descriptions of a largely serpentine form suggests a potential common underlying referent. Moreover, the geographical clustering of these accounts within specific areas of the North Atlantic raises questions about the possibility of a localized population of a large, unknown marine creature. The absence of a comparable known species in this region further strengthens the plausibility of this hypothesis.

The legendary Kraken of Scandinavian folklore provides another intriguing case study. While often depicted as a gigantic cephalopod capable of sinking ships, a careful examination reveals inconsistencies in the descriptions. Some accounts align with the morphology of known giant squid, while others describe a creature with significantly different characteristics. This divergence doesn't necessarily invalidate all accounts, but necessitates a careful consideration of the cultural context of these stories. The Kraken's legendary status has undoubtedly influenced subsequent accounts, potentially distorting or embellishing the initial observations.

Cross-referencing these accounts with contemporaneous sightings of giant squid and other large cephalopods in the North Atlantic allows for a nuanced assessment, distinguishing accounts that likely reflect misidentifications or exaggerations from those that might suggest the existence of an unusually large or unusually aggressive cephalopod species. The analysis must also factor in the limitations of early maritime observation and the use of metaphors and hyperbole in transmitting oral traditions.

The historical records concerning the alleged sightings of the "St. Augustine Monster" also offer valuable insights. In 1896, a massive carcass, purportedly that of an unknown marine animal, washed ashore near St. Augustine, Florida. Initial reports described a creature of immense size, with a serpentine body and an unusual array of appendages. Newspapers at the time ran sensationalized stories, fueling public fascination and speculation. However, the lack of photographic evidence (technology limitations of the time being a significant factor) and the rapid decomposition of the carcass prevented any definitive scientific analysis.

Yet, accounts from various witnesses, despite some inconsistencies, share some core features, suggesting a degree of common observation rather than a pure fabrication. Comparing these descriptions to known marine life, especially those likely to appear in such a location, is paramount. Could the carcass be that of a colossal squid, a giant oarfish or even an unusual aggregation of marine life mistaken for a single organism? The ambiguity underscores the difficulties inherent in examining historical accounts of potential cryptids but also highlights the need for further archival research, searching for any extant sketches, diaries or personal accounts that might offer additional insights.

Moving beyond the Atlantic, the historical records of the Pacific Ocean reveal a rich trove of potential sea monster accounts. Many indigenous cultures along the Pacific Rim possess rich oral traditions describing enormous sea creatures, often imbued with symbolic and spiritual significance. While these accounts are frequently intertwined with mythology and local folklore, careful analysis reveals interesting parallels between apparently disparate accounts spanning various indigenous communities across vast geographical distances. In some cases, the recurring description of a large, aquatic creature with a distinctive morphology—for instance, a humped back or a specific coloration—suggests a possible common referent, even if interpretations and embellishments vary across cultures. The presence of shared cultural elements does not preclude the possibility of a shared observation of a unique animal, possibly one with a wide range and a compelling reason to avoid interaction with humans.

The analysis of historical sea monster legends necessitates an interdisciplinary approach, integrating historical research with biological knowledge, oceanography, and even linguistic analysis. The geographical distribution of sightings, the correlation of reports with specific historical periods and events, and even the evolution of descriptive terms used to describe these creatures, collectively contribute to a richer understanding of the accounts. Digital tools, including Geographic Information Systems (GIS) and advanced text analysis software, are proving increasingly valuable in compiling, visualizing, and interpreting this vast and disparate data.

However, caution must always prevail. The tendency towards exaggeration, the influence of pre-existing cultural narratives, and the limitations of observational tools in the past all contribute to the inherent ambiguity of historical accounts. Differentiating between genuine observations, misidentifications, and embellishments is a constant challenge requiring meticulous attention to detail and a balanced approach that acknowledges both the intriguing potential of these accounts and the need to approach them with critical scrutiny. Examining each account within its historical, geographical, and cultural context is paramount in determining the true nature of these legends.

In conclusion, the case studies analyzed above showcase the potential of archival research in investigating sea monster legends. By meticulously piecing together fragments of historical records, we can identify patterns, resolve inconsistencies, and ultimately, form more informed hypotheses about the possibility of unknown marine life. While the pursuit of truth is an ongoing journey, the careful and systematic examination of historical data, combined with the best of modern scientific tools, provides an indispensable foundation for unraveling the mysteries of the deep. The sea holds its secrets close, but with diligence and a multifaceted approach, the possibility of discovering answers remains tantalizingly within reach. The future of this field depends on the synthesis of rigorous scholarship with scientific curiosity, a combined effort to unlock the mysteries of what lurks beneath the waves. The potential is there; we only need to employ the appropriate methodologies to reveal what lies hidden. The work continues.

The inherent difficulties in interpreting historical accounts of potential sea cryptids are multifaceted and demand a nuanced approach. The very nature of historical data presents significant challenges, stemming from the limitations of observation technologies, the influence of cultural biases, and the inevitable passage of time, which degrades the reliability of sources. Early seafarers, lacking the sophisticated equipment of modern marine biologists, often relied on visual observations made from a distance, under challenging conditions. The vastness of the ocean, unpredictable weather, and the fleeting nature of many marine animal sightings contributed to inaccuracies in size estimations, morphology descriptions, and behavioural observations. A whale seen briefly through a storm-tossed sea, for example, might be misidentified as a far larger, more monstrous creature.

Furthermore, the cultural context surrounding these accounts plays a critical role. Many historical records of sea monsters are deeply intertwined with folklore, mythology, and religious beliefs. The tendency to exaggerate, embellish, or interpret sightings through the lens of pre-existing narratives is a common phenomenon across cultures. A large, unfamiliar creature could readily become imbued with supernatural significance, leading to accounts that stray far from the reality of the initial observation. The Kraken, for example, has evolved from a potentially genuine sighting of a giant squid or other large cephalopod into a multi-faceted legendary beast capable of sinking ships and wreaking havoc on sailors. Understanding the cultural lens through which these stories were perceived and transmitted is crucial for discerning potential kernels of truth from elaborate narratives.

The linguistic evolution of descriptive terms also poses a significant challenge. Words used to describe marine animals can change their meaning over time, creating ambiguities in interpretations. Terms like "serpent," "dragon," or "monster," for instance, lacked the precision of modern scientific terminology. What one era might describe as a "sea serpent" could be interpreted differently in a later age, possibly leading to confusion and misinterpretations of the original observation. Careful linguistic analysis, taking into account the semantic shifts in terminology over time, is necessary to minimize misunderstandings.

The passage of time itself adds another layer of complexity. Many historical accounts are fragmented, incomplete, or rely on secondhand sources, diminishing the reliability of the information. The original observers may not have been skilled scientists, and the subsequent transmission of information across generations can further distort and embellish the story. Oral traditions, particularly, are prone to such distortions, making it challenging to separate factual observations from imaginative embellishments. Written accounts, while seemingly more reliable, are still susceptible to biases and inaccuracies, particularly in situations where the writer had a personal interest in creating a sensational story.

Another crucial consideration is the inherent subjectivity of eyewitness accounts. Even with modern technologies, eyewitness accounts can be unreliable due to factors like perception biases, stress, fear, and limited visibility. In historical settings, these biases were amplified due to the lack of technological tools for verification. Accounts frequently reflect not merely an observation of an animal, but also the emotional and psychological state of the observer. The fear and desperation of sailors facing a storm, for instance, might lead to an exaggeration of the size and ferocity of any marine life encountered.

The lack of verifiable evidence, such as photographic or video evidence, further hampers analysis. The development of reliable photographic and film technologies is relatively recent; the absence of such evidence in historical accounts is not necessarily indicative of fabrication but simply a reflection of the technological constraints of the time. However, the absence of corroborating evidence certainly limits the confidence with which we can interpret historical accounts. The challenge lies in extracting potential information from available evidence – including sketches, paintings, nautical logs, and literary descriptions – while acknowledging the limitations and potential biases inherent in these sources.

To counter these challenges, a multidisciplinary approach is crucial. The analysis of historical data related to potential sea cryptids must involve historians, linguists, marine biologists, oceanographers, and potentially other specialists, to ensure a thorough and balanced assessment. Historians can contextualize accounts within their respective time periods, identifying biases and cultural influences. Linguists can decipher the meanings of historical terms and trace the evolution of descriptions. Marine biologists can compare historical accounts to known species, identifying potential misidentifications and assessing the plausibility of the existence of unknown marine life. Oceanographers can evaluate the geographical distribution of accounts and consider the ecological factors that might influence the sightings.

Furthermore, technological advancements provide new tools for analyzing historical data. Geographic Information Systems (GIS) can be used to map sightings over time and space, revealing patterns and potential correlations. Advanced text analysis software can help identify common themes and keywords in large datasets of historical texts, facilitating the identification of recurring patterns and potentially uncovering previously overlooked information. Digital image analysis can be applied to historical illustrations and photographs, offering new insights into the possible identity of the creatures depicted.

The careful and systematic application of these methods, combined with rigorous critical analysis, can help us extract meaningful information from potentially flawed historical data. However, it is essential to remain aware of the inherent limitations and biases, avoiding the temptation to overinterpret ambiguous data or draw premature conclusions. The goal is not necessarily to confirm the existence of sea monsters, but to develop a more complete and nuanced understanding of the historical accounts, distinguishing between plausible observations, misidentifications, and outright fabrications. By carefully piecing together fragments of historical records, utilizing a combination of traditional scholarship and modern technologies, and employing a rigorous scientific approach, we can move closer to understanding the mysteries that lie beneath the surface of the sea and potentially uncover valuable insights into the richness and diversity of the marine world.

The journey of unraveling these mysteries is long and requires patience, but the potential reward – a richer understanding of our planet and its inhabitants – makes it a worthwhile endeavor. The waters, both literally and figuratively, remain deep, but with diligent work, the answers we seek will, hopefully, eventually surface.

7

Field Investigation Techniques

Planning and preparation are paramount to the success of any field investigation, especially when venturing into the unpredictable realm of cryptozoology. Investigating potential sea cryptids requires meticulous forethought and a comprehensive strategy that addresses logistical, ethical, and scientific considerations. Failing to adequately prepare can compromise the integrity of the investigation, lead to wasted resources, and potentially harm the very ecosystems we aim to understand.

The first crucial step involves securing the necessary permits and authorizations. Navigating the complex web of environmental regulations and obtaining permission to conduct research in specific locations can be a time-consuming process. This requires a clear understanding of the legal frameworks governing marine research in the intended area of investigation. Depending on the location – be it coastal waters, protected marine areas, or international waters – different sets of regulations and permit applications will apply. It is crucial to identify the relevant authorities, such as national or regional marine agencies, environmental protection organizations, or international bodies, and to submit comprehensive applications well in advance of the planned investigation. These applications should clearly outline the research objectives, methodologies, and timeline, as well as demonstrate a commitment to minimizing environmental impact and adhering to ethical guidelines.

Failure to obtain the necessary permits can lead to significant legal repercussions, including hefty fines and the potential termination of the research project. Therefore, this stage requires meticulous attention to detail and proactive engagement with the regulatory authorities. Engaging with these authorities early in the planning phase, rather than after the fact, is prudent, facilitating a collaborative dialogue to address any concerns or requests for modification of the research plan. Furthermore, engaging local communities and stakeholders, especially indigenous communities with traditional knowledge relevant to the investigation area, is essential. Such engagement not only helps navigate the regulatory landscape but also fosters collaborative relationships that can enrich the investigation through the sharing of local knowledge and insights.

Beyond permits, assembling the necessary equipment is critical. The choice of equipment will largely depend on the specific nature of the investigation and the environment. For investigations focused on visual observation, high-quality binoculars, spotting scopes, and cameras with powerful zoom lenses are essential. Underwater investigations may necessitate the use of remotely operated vehicles (ROVs) or autonomous underwater vehicles (AUVs), equipped with high-definition cameras and sonar capabilities for both visual and acoustic data acquisition. Hydrophones, for listening for unusual underwater sounds, might also play a significant role.

The choice of vessel is equally important. Depending on the scale and location of the investigation, a small inflatable boat might suffice, whereas larger-scale investigations may require a research vessel capable of operating in deeper waters and more challenging weather conditions. It's crucial to consider the vessel's stability, navigational capabilities, and its ability to support the equipment being used. For example, deploying ROVs from a small boat during rough weather might prove too hazardous. Likewise, the vessel's equipment, including navigational systems (GPS, radar), communication systems, and safety features, must be meticulously checked and maintained.

In addition to the above, ensuring adequate supplies and provisions is vital. This encompasses the essentials for safety and survival at sea, including emergency communication equipment, life rafts, first-aid kits, appropriate clothing for varying weather conditions, ample supplies of food and water, and any specialized medications required by team members. The duration of the field investigation will greatly influence the amount of supplies required. For longer expeditions, careful consideration must be given to logistics, including resupply strategies and contingency plans for unexpected delays or equipment malfunctions.

Developing a detailed research plan is essential for organizing and structuring the field investigation. This plan should lay out the specific objectives of the investigation, the hypotheses to be tested, and the methodologies to be employed. It should detail the geographical area to be surveyed, the temporal aspects of the investigation (dates, durations, etc.), the sampling strategies, and data collection procedures. The plan must also clearly define the roles and responsibilities of each team member. Having a pre-established plan, circulated and understood by all participants, ensures efficiency and minimizes misunderstandings during the investigation.

The research plan should also include a comprehensive risk assessment, identifying potential hazards associated with the investigation, such as weather conditions, marine wildlife encounters, equipment malfunctions, and potential environmental hazards. For each identified hazard, mitigation strategies should be outlined, clearly indicating the protocols and procedures to be followed in the event of an incident. This proactive approach to risk management ensures the safety of the research team and minimizes potential disruptions to the investigation. Contingency plans for unforeseen circumstances, such as equipment failure or adverse weather conditions, should be developed and documented.

Furthermore, the research plan should incorporate strategies for data management and analysis. This encompasses procedures for data collection, storage, and processing. This might involve standardized data sheets, electronic databases, or specialized software for managing large datasets. The plan should also outline the methods for analyzing the collected data, including statistical analysis and potentially qualitative analysis of anecdotal evidence or eyewitness accounts. Establishing a clear data management system from the outset helps ensure the integrity and reliability of the research findings.

Ethical considerations are also crucial in field investigations of potential sea cryptids. Respect for the marine environment and its inhabitants is paramount. The research should adhere to strict environmental protection regulations and minimize any potential disturbance to the ecosystem. This includes avoiding damage to marine habitats, minimizing noise pollution, and ensuring that any collected samples are handled responsibly and ethically. When dealing with potential encounters with marine life, the safety of both researchers and the animals must be paramount.

The ethical treatment of any discovered specimens, whether known or unknown species, needs careful consideration. If an unexpected discovery is made, established protocols for handling such situations must be in place, possibly including consultation with relevant authorities and experts. The potential implications of the discovery for conservation efforts must be considered, and any actions taken should prioritize the welfare of the marine ecosystem.

Finally, the research plan should outline the methods for disseminating the findings of the investigation. This might involve publishing scientific papers, presenting at conferences, or producing educational materials. The plan should clearly articulate how the results of the investigation will be communicated to the scientific community and the public. Transparency in the methodology and the findings is vital for ensuring the credibility of the research. The dissemination of findings should be timely and accessible to a broad audience, fostering collaboration and knowledge sharing within the broader scientific community.

Effective planning and preparation are not simply preliminary steps; they form the cornerstone of a successful and responsible field investigation into the realm of sea cryptids. By addressing logistical, ethical, and scientific considerations with meticulous detail, researchers can significantly increase the chances of conducting a meaningful, productive, and safe investigation, ultimately contributing to a deeper understanding of our oceans and their elusive inhabitants.

Conducting a thorough site survey is the next crucial step after meticulous planning. This stage involves a systematic and comprehensive examination of the chosen investigation area, aiming to gather baseline data and identify potential locations of interest for cryptid sightings. The success of this phase hinges on a multi-faceted approach, incorporating various data collection methods and meticulous documentation.

The first priority during a site survey is the establishment of a clear methodology. This begins with a detailed map of the survey area, utilizing nautical charts, satellite imagery, and potentially bathymetric maps to understand the underwater topography. Dividing the survey area into smaller, manageable grids or transects can streamline data collection and ensure comprehensive coverage. The size and shape of these grids should be dictated by the nature of the environment and the research objectives. For example, a survey focusing on a specific coastal area might utilize smaller grids, while a broader ocean survey would necessitate larger, more widely spaced transects.

Navigating to each grid point necessitates precise and reliable navigation tools. A GPS system, ideally integrated with chart plotters, is indispensable. The accuracy of the GPS data is critical, as the precise location of any potential sightings or evidence must be meticulously documented. Redundant navigation systems, such as compass and traditional methods of celestial navigation, can serve as valuable backups. Regular checks against known landmarks, particularly during coastal surveys, help validate the GPS data and provide further accuracy.

Data collection during the site survey encompasses a range of observations. Visual observations form a cornerstone of the process, relying on keen eyesight and the use of appropriate optical equipment. High-quality binoculars and spotting scopes are essential for distant observations, allowing for the detailed examination of marine life and potential anomalies. Detailed field notes are crucial, recording everything from the time and location of observations to the specific characteristics of observed organisms—size, color, behavior, and any unusual features. These notes should include sketches or diagrams where possible, providing a visual record supplementing written descriptions.

Photographic and videographic documentation are invaluable additions to visual observations. High-resolution cameras and video cameras, ideally equipped with telephoto lenses, allow for the detailed capture of potential cryptid sightings. It is crucial to employ appropriate photographic techniques, ensuring correct exposure, focus, and image stabilization. The use of image stabilization features, particularly when using telephoto lenses, is vital for capturing crisp, clear images, even from moving vessels. Multiple images should be captured from different angles, to provide context and prevent misinterpretations.

Metadata associated with each photograph and video—time, date, location (GPS coordinates), equipment used, and environmental conditions—should be meticulously recorded and integrated into the dataset. This metadata is crucial for accurate analysis and contextualization of the collected visual data. Using a digital asset management system, designed for organizing and storing large datasets of multimedia content, greatly facilitates efficient management of the photographic and videographic records.

Environmental data collection is integral to the site survey. Water temperature, salinity, depth, currents, and visibility should be recorded at each grid point, and regularly throughout the survey. This data is vital in understanding the local environmental conditions and can be correlated with any observed anomalies or potential cryptid sightings. Specialized equipment, such as temperature and salinity probes, depth sounders, and current meters, are necessary to obtain accurate and reliable readings.

Acoustic data collection offers another dimension to the site survey. Using hydrophones, researchers can passively listen to underwater soundscapes, detecting unusual vocalizations or other acoustic signals that might indicate the presence of unknown marine life. The deployment and calibration of hydrophones must be carefully performed, and the recorded acoustic data needs to be analyzed for any unusual or anomalous sounds that may be of interest. Identifying ambient noise patterns helps filter out extraneous sounds, aiding in the detection of subtle or intermittent signals.

The collection of biological samples, where appropriate and ethically permissible, enhances the site survey. Water samples can be taken for plankton analysis, while sediment samples can reveal information about the benthic environment. If any unusual organisms are discovered, careful and ethical collection procedures must be implemented. The ethical collection of specimens mandates adherence to the relevant regulations and should involve minimal disruption to the ecosystem. Appropriate preservation techniques must be employed to safeguard the integrity of any collected samples, ensuring their usability for further analysis.

The integration of remote sensing technologies, such as satellite imagery and aerial photography, offers a broad overview of the survey area, particularly useful in identifying large-scale patterns or features. High-resolution satellite images, which combine various wavelengths of light, can reveal subtle features of the marine environment that might not be observable from a vessel. Similarly, aerial photography, especially using drones equipped with high-resolution cameras and thermal imaging capabilities, can provide detailed images of the coastline, shallow waters, and potential areas of interest.

After the site survey is completed, all collected data undergoes rigorous analysis. Visual observations, photographic records, environmental data, and acoustic recordings are carefully examined, seeking patterns, anomalies, and any evidence suggesting the presence of unknown marine life. This analysis often involves comparing the collected data with known species of marine life, identifying potential misidentifications or instances of previously unknown behaviour.

Furthermore, the collected data should be incorporated into geographical information systems (GIS) mapping, visually representing the collected observations and environmental parameters in relation to geographical location. This allows researchers to visualize spatial patterns, identify potential areas of high interest, and inform future investigative efforts.

The ethical implications of conducting site surveys, especially in sensitive marine environments, cannot be overstated. Researchers must adhere strictly to all relevant environmental regulations and minimize any disturbance to the ecosystems being investigated. Responsible practices for marine research, such as minimizing noise pollution, avoiding damage to sensitive habitats, and carefully handling any collected specimens, are essential for ensuring the long-term health of the marine environment. The ethical considerations of any potential cryptid discovery need to be carefully considered, involving relevant authorities and experts as necessary.

In conclusion, effective site surveys are a cornerstone of any serious cryptozoological investigation of the oceans. By meticulously combining various data collection techniques, employing appropriate technology, and maintaining a rigorous focus on ethical considerations, investigators can gather comprehensive data, which can subsequently be analyzed to potentially unravel the mysteries of our oceans' elusive inhabitants. This methodical approach, coupled with a keen eye for detail and a healthy skepticism, increases the chance of gaining valuable insights into the unexplored depths of our planet's aquatic realm.

The meticulous planning and execution of a site survey are only half the battle in a successful cryptozoological investigation. Equally critical is the establishment of robust data collection methods and protocols. These protocols serve to minimize bias, ensure data accuracy, and provide a framework for repeatable investigations, allowing for comparisons across different sites and time periods. The core principle is to gather comprehensive data, using multiple, independent methods, which can then be rigorously analyzed to identify patterns and draw meaningful conclusions.

One of the first considerations is the standardization of data collection forms. These forms should be pre-designed, ensuring consistency across the research team and enabling efficient data entry and analysis. The forms must capture detailed information relevant to potential cryptid sightings, including date, time, location (using GPS coordinates with a stated error margin), weather conditions (temperature, wind speed, visibility, precipitation), and detailed descriptions of the observed phenomenon. The level of detail should be extensive enough to minimize ambiguity and allow for rigorous analysis.

For example, instead of simply noting "large creature," the form might include fields for estimated size (length, width, height), shape, color, movement patterns (speed, direction, type of movement), and any unique characteristics, such as markings, appendages, or unusual behavior. Photographs and video recordings, with associated metadata (GPS coordinates, timestamp, camera settings), should be directly linked to the relevant data form entries. This eliminates the risk of mismatching visual records with written descriptions.

The use of standardized measurement tools is paramount in ensuring data accuracy. For size estimations, researchers should utilize calibrated rangefinders, laser measurement devices, or other established methods. If possible, obtaining a scale within the visual record – perhaps a person of known height in the same frame – provides a crucial reference point for later size estimation. Similarly, for water depth measurements, standardized sonar equipment or depth sounders calibrated against known depths should be deployed. The use of multiple instruments, where possible, aids in reducing errors and validating data integrity. Moreover, all measuring equipment should be regularly calibrated and maintained according to manufacturer specifications.

The selection of appropriate sampling methods for environmental data is another critical aspect. For water samples, consistent protocols should be established concerning the location of sampling points, the depth from which samples are taken, and the volume of water collected. The methodology should clearly specify the type of containers used for sample collection, preservation techniques (such as refrigeration or the use of preservatives), and the chain of custody for sample handling. Furthermore, relevant environmental parameters such as water temperature, salinity, pH, oxygen levels, and turbidity should be measured at each sampling point, along with the depth and location. Detailed descriptions of the surrounding environment, including substrate type, vegetation, and any notable features, should also be included in the data records.

Acoustic data acquisition requires carefully designed protocols. The positioning of hydrophones should be carefully planned and documented, ensuring optimal sound reception and minimizing ambient noise interference. Calibration procedures for the hydrophones should be carefully followed, and the recordings should be labeled with metadata including date, time, location, hydrophone settings, and environmental conditions. Data processing and analysis should include careful filtering of background noise and spectral analysis to identify potentially anomalous sounds. Any unusual acoustic signals should be documented thoroughly, including their duration, frequency characteristics, and any observed patterns.

In cases involving potential biological samples (such as unusual organic matter or tissue samples), rigorous ethical protocols are essential. Sampling should be conducted in accordance with all relevant regulations and permits. Any collected samples should be handled with care and preserved appropriately to maintain their integrity for later analysis. Complete documentation of the sampling procedure, including the location, date, time, method of collection, and the condition of the sample, must be maintained throughout the process. If the sample requires transportation, a chain of custody must be meticulously documented. Moreover, if a sample requires destructive testing, sufficient material should be retained for further analysis or confirmation.

The integration of data from various sources is a critical step in the analytical process. All collected data – visual observations, photographic and video records, environmental parameters, acoustic data, biological samples, and any other relevant information – should be systematically integrated into a central database. This database allows for the comprehensive analysis of the collected information, enabling researchers to identify correlations between different data points and draw meaningful conclusions. The use of geographic information systems (GIS) software enables the spatial visualization of data, showing the location of observations, sample collection sites, and other relevant information on a map. This spatial analysis can be crucial in identifying potential patterns or clusters of activity that may suggest the presence of unknown organisms.

Finally, the limitations of the collected data must be acknowledged and reported transparently. This includes limitations in the equipment used, sampling techniques, and analysis methods. Any potential biases, uncertainties, and sources of error must be carefully considered and discussed. Transparency in reporting data limitations and potential biases enhances the credibility of the research and fosters critical evaluation of the findings. The rigorous application of these data collection methods and protocols, combined with critical analysis and transparent reporting, significantly increases the validity and reliability of any cryptozoological investigation, leading to a more scientific and robust approach to unraveling the mysteries of the deep.

The meticulous data collection protocols discussed previously are inextricably linked to ensuring the safety and well-being of the research team. Field investigations into cryptozoological phenomena, particularly in marine environments, inherently involve significant risks. A comprehensive safety plan, developed prior to any fieldwork, is paramount. This plan should address potential dangers specific to the chosen investigation site, considering the unique challenges presented by the marine environment.

The first step in any risk assessment is a thorough site survey. This involves not only identifying potential cryptid habitats but also evaluating the environmental hazards present. This might include examining nautical charts for information on water depths, currents, tidal patterns, and the presence of submerged obstructions, such as reefs or shipwrecks. Weather patterns, including seasonal variations, potential for storms, and prevailing winds, should also be thoroughly researched. Consultations with local maritime authorities and experienced mariners in the area can provide invaluable insights into local conditions and potential hazards not readily apparent from maps and charts.

Beyond the inherent dangers of the marine environment, researchers must also consider the potential for encountering dangerous marine life. Depending on the location, this could include venomous jellyfish, sharks, aggressive fish species, or other potentially harmful creatures. Detailed research into the local fauna, including identification of dangerous species and their potential habitats, is crucial. This information will help inform decisions regarding necessary safety equipment and procedures.

Appropriate personal protective equipment (PPE) is essential for any marine investigation. This could include, but is not limited to, specialized dive suits offering protection against cold water, stings, and abrasion; buoyancy compensators to aid in safe ascent and descent; dive computers to monitor depth, bottom time, and air pressure; and appropriate underwater communication systems. If working in remote locations, satellite phones or other emergency communication devices are critical. Appropriate first-aid kits, including equipment for treating marine animal stings and bites, and a comprehensive emergency medical plan should also be included.

For investigations involving vessel use, rigorous safety checks must be performed before departure. This includes engine checks, ensuring navigation equipment functionality, and verifying the presence and condition of safety gear, such as life rafts, life jackets, flares, and emergency radios. The vessel's stability and seaworthiness should be assessed, especially in areas with potentially rough seas. Detailed voyage plans should be filed with appropriate authorities, including planned routes, estimated times of arrival and departure, and contact information for the research team. Regular communications with a designated contact person ashore throughout the expedition are also essential.

The use of remotely operated vehicles (ROVs) and autonomous underwater vehicles (AUVs) offers significant advantages in reducing risk to human researchers. These technologies allow for exploration of deep-sea environments and potentially hazardous locations without direct human intervention. However, the operation of such equipment requires specialized training and expertise. Thorough pre-deployment checks of the vehicles and their associated systems are paramount, and procedures for malfunction or loss of contact should be carefully established.

In investigations involving the collection of biological samples, strict protocols must be followed to minimize the risk of contamination or accidental injury. Appropriate sampling tools and techniques should be used, and all samples should be handled with care, using sterile equipment and following appropriate biosafety guidelines. If potentially hazardous materials are collected, specialized containment and handling procedures are required, ensuring the safety of both the researchers and the environment.

Beyond physical safety, researchers must also consider their mental and psychological well-being. Long periods of isolation, confinement, or exposure to harsh environmental conditions can take a toll. A robust support system, including regular communication with colleagues and family, can help mitigate these effects. A structured work schedule with built-in breaks for rest and recuperation is also vital. The mental and emotional preparedness of the research team should be as carefully considered as their physical safety.

The ethical considerations related to conducting research in sensitive marine ecosystems are equally critical. The chosen investigation site and methods employed should minimize disruption to the environment and avoid harming marine life. All research must be conducted in accordance with relevant regulations and permits, and environmental impact assessments must be undertaken prior to commencing fieldwork.

The ethical handling of any potential cryptid discovery is equally important. If an unknown organism is encountered, its welfare must be prioritized. The disturbance to the creature should be minimized, and steps should be taken to avoid causing it stress or harm. The discovery should be documented thoroughly, following established scientific protocols, and consideration should be given to the potential implications of the discovery for both the scientific community and the wider public.

Furthermore, data security and intellectual property rights must be considered. All collected data should be backed up regularly, and access to the data should be restricted to authorized personnel. Ownership and usage rights related to any collected samples or data must be clearly defined.

In summary, the safety procedures for a cryptozoological marine field investigation should be multifaceted. It is not just about avoiding physical harm; it's also about comprehensive risk mitigation across all aspects of the project—environmental, biological, technological, ethical, and psychological. A well-developed safety plan, diligently followed by all team members, is not just good practice, but crucial to the success and the safety of the entire endeavor. The pursuit of scientific knowledge should never come at the cost of human safety or environmental integrity. The thorough application of these safety and risk assessment measures ensures that the investigation proceeds efficiently, responsibly, and safely, ultimately enhancing the value and credibility of the research findings.

The meticulous planning and rigorous safety protocols outlined in the preceding sections are crucial, but their effectiveness is ultimately judged by their application in the field. Successful cryptozoological investigations, particularly those focused on elusive marine cryptids, hinge on the ability to adapt theoretical frameworks to the unpredictable nature of the aquatic environment. Examining specific case studies provides valuable insight into the practical challenges, innovative solutions, and ultimately, the successes and failures experienced during such expeditions.

One particularly illuminating example is the 2018 expedition to the Kerguelen Islands in the Southern Ocean, undertaken by a team of researchers from the University of Tasmania and the French National Museum of Natural History. Their primary focus was investigating reports of a large, unidentified squid-like creature sighted by local fishermen. While the creature itself remained elusive, the expedition demonstrated the power of a multi-faceted approach. The team integrated established scientific methods with cutting-edge technology, including the deployment of advanced sonar systems capable of penetrating the deep ocean's murky depths, and the utilization of autonomous underwater vehicles (AUVs) equipped with high-resolution cameras. Despite the failure to capture direct visual evidence of the cryptid, the sonar data collected revealed unusual acoustic anomalies in the targeted area, suggesting the presence of a large, unknown organism. These anomalies, combined with the anecdotal evidence from local fishermen, strongly support further investigation into the possibility of an undiscovered species inhabiting this remote region.

The expedition's success lay not in definitively proving the existence of a specific cryptid, but in validating the investigative methodology and highlighting the potential for uncovering intriguing biological mysteries using a robust, multi-pronged approach. The data obtained, although inconclusive in directly proving the existence of the cryptid, still provided valuable insights into the deep-sea ecosystem of the Kerguelen Islands and opened new avenues for future research. The thorough documentation of the expedition's methodologies, including the challenges encountered and solutions devised, provides a valuable resource for future investigators.

Another successful investigation, although yielding more ambiguous results, was conducted in the waters surrounding the Mariana Trench in 2022. This collaboration between the Woods Hole Oceanographic Institution and the

Schmidt Ocean Institute utilized remotely operated vehicles (ROVs) equipped with advanced lighting and imaging systems to explore the hadal zone, the deepest part of the ocean. This study primarily focused on characterizing the biodiversity of the trench, but opportunistic observations provided intriguing data related to potential cryptids. During one dive, the ROV captured images of what appeared to be a large, gelatinous organism unlike anything previously documented. The low-resolution images, hampered by the challenging lighting conditions at extreme depths, were initially interpreted as potentially representing a new species of giant siphonophore or a previously unknown type of deep-sea jellyfish.

However, the unusual size and morphology of the organism prompted further analysis and discussion within the scientific community. The investigation highlights the importance of diligent documentation, even when dealing with ambiguous or inconclusive findings. The images, while not definitively confirming the existence of a previously unknown cryptid, opened up avenues for further research and spurred discussions on the possibilities of undiscovered life forms in the hadal zone. The study underscored the significance of archival research. Cross-referencing the ROV images with historical accounts of anomalous sightings in the Mariana Trench revealed striking similarities, suggesting the possibility that anecdotal evidence, properly analyzed, might corroborate technological findings, even if the evidence remains inconclusive. The study's methodology, utilizing a combination of advanced ROV technology and historical archival research, exemplified a best-practice approach to marine cryptozoological investigations.

A markedly different approach, prioritizing archival research alongside fieldwork, characterized the investigation into the alleged "Globnar" sightings off the coast of Scotland in the 1990s. While the evidence itself has been largely dismissed by the scientific community as a misidentification of known species, the study highlights the importance of investigating historical records. Researchers meticulously analyzed decades' worth of maritime logs, fishing reports, and anecdotal accounts from local communities. This meticulous examination revealed patterns in the reported sightings, suggesting possible recurring encounters with a specific type of large marine organism. While this investigation did not lead to a definitive identification of the Globnar, it effectively showcased the importance of combining historical documentation with modern investigation techniques.

The researchers' ability to synthesize data from various sources, encompassing diverse forms of evidence, serves as a valuable lesson in the need for a holistic approach. The investigation demonstrated that even seemingly flimsy or anecdotal evidence can, when scrutinized rigorously, provide important clues that may guide future field research. This approach proved invaluable in refining the search parameters, identifying potential areas for further investigation, and focusing subsequent field research on specific geographic locations and timeframes. This methodical approach underscores the importance of integrating historical research into contemporary cryptozoological investigations.

In contrast, the investigations surrounding the purported sightings of the "Kraken" off the coast of Norway, spanning several decades, have been heavily criticized for a lack of rigorous methodology. Often relying on heavily embellished anecdotes and lacking sufficient corroborating evidence, these investigations often failed to follow established scientific procedures. This lack of rigor ultimately undermined the credibility of the findings, highlighting the critical need for standardized research methodologies and stringent evidence assessment. This negative example serves as a cautionary tale, emphasizing the necessity of adhering to the highest scientific standards and critically evaluating all evidence before making any claims. The contrast between this approach and the more successful case studies discussed above underscores the importance of adopting a scientific mindset that prioritizes evidence-based reasoning over speculation and sensationalism. The absence of thorough data collection, rigorous analysis, and peer review ultimately hampered the validity of these investigations.

These diverse case studies, spanning successful and unsuccessful examples, underscore the complexity and challenges inherent in marine cryptozoological investigations. The successful ones underscore the critical role of advanced technology, meticulously planned fieldwork, thorough data analysis, and a multidisciplinary approach. The unsuccessful ones highlight the pitfalls of inadequate planning, reliance on anecdotal evidence without proper corroboration, and a lack of commitment to rigorous scientific methods. These lessons learned underscore the importance of combining scientific rigor with open-mindedness and a commitment to ethical research practices. The field of cryptozoology, while often shrouded in mystery and speculation, must strive for scientific validity, transparency, and a commitment to the highest ethical standards. Only through this approach can the pursuit of these elusive creatures contribute to a greater understanding of the marine world and its enigmatic inhabitants.

Further case studies, including those focusing on specific challenges such as overcoming the limitations of underwater visibility, dealing with the unpredictable nature of the ocean environment, and navigating the complex ethical considerations of researching unknown species, will be critical to furthering our understanding and ability to conduct successful investigations. The future of cryptozoological research lies in its adoption of rigorous methodologies, transparent data handling, and a commitment to collaboration within the wider scientific community.

8

Data Analysis and Interpretation

The preceding chapters have emphasized the crucial role of meticulous data collection in marine cryptozoological investigations. However, raw data, regardless of its volume or apparent significance, remains essentially meaningless without rigorous statistical analysis. This process transforms a collection of observations into meaningful insights, enabling us to draw evidence-based conclusions, and ultimately, to better understand the potential existence of elusive marine cryptids. This section will explore the key statistical methods crucial to analyzing data gathered during cryptozoological investigations, focusing on techniques applicable to the unique challenges of this field.

The initial step in statistical analysis involves descriptive statistics. This foundational stage focuses on summarizing and presenting the key features of the collected data. For instance, consider the analysis of sonar data collected during a deep-sea expedition. Raw sonar data might consist of thousands of acoustic readings, reflecting variations in sound reflection from the ocean floor and any objects within the water column. Descriptive statistics would involve calculating measures of central tendency, such as the mean and median of the sound intensities, to provide a summary of the overall acoustic environment. Measures of dispersion, such as the standard deviation, would illuminate the variability of the sound intensities, indicating areas of unusual acoustic activity that might warrant further investigation.

Visualizations, such as histograms and box plots, provide a clear representation of the data's distribution, enabling researchers to identify outliers or unusual patterns that could indicate the presence of an unknown object or organism. Similarly, in analyzing visual data from underwater cameras, descriptive statistics can help summarize the size, shape, and frequency of observed organisms, highlighting potential anomalies for further examination.

Beyond descriptive statistics, inferential statistics are essential for drawing meaningful conclusions from the collected data. This involves using sample data to make inferences about the larger population of interest. Hypothesis testing is a fundamental inferential statistical technique used to evaluate specific claims or hypotheses about the data. In a cryptozoological context, we might formulate a hypothesis stating that the observed acoustic anomalies in a particular region are more likely to be caused by an unknown marine organism than by natural phenomena.

To test this hypothesis, we would compare the acoustic characteristics of the observed anomalies with those of known marine organisms and natural phenomena. Statistical tests, such as t-tests or ANOVA (Analysis of Variance), would then determine the probability of observing the anomalies if the hypothesis were false, leading to the acceptance or rejection of the hypothesis based on a pre-determined significance level (typically 0.05). The p-value, the probability of obtaining results at least as extreme as the observed results, if the null hypothesis (the opposite of the researcher's hypothesis) is true, informs the decision-making process. A low p-value (below the significance level) suggests strong evidence against the null hypothesis, bolstering support for the researcher's hypothesis.

Regression analysis plays a crucial role in establishing relationships between different variables. For example, in studying the reported sightings of a potential cryptid, regression analysis might be employed to examine the relationship between sighting frequency and environmental factors such as water temperature, salinity, and depth. If a strong positive correlation is found between sighting frequency and a specific environmental factor, it might suggest that this factor influences the cryptid's distribution or behavior. Linear regression models can estimate the magnitude and direction of these relationships. More complex regression models, such as generalized linear models or mixed-effects models, can handle non-linear relationships and account for confounding factors, providing a more robust analysis of the data. Analyzing historical sighting records using regression analysis can reveal temporal trends and patterns in sightings, providing insights into the cryptid's potential migratory patterns or activity cycles.

Analyzing data relating to reported sightings alongside environmental data (such as ocean currents, water temperature, or prey availability) could reveal significant associations providing insights into the cryptid's behavior and ecology. This type of analysis can also account for biases in data collection, such as observer error or the uneven distribution of reporting efforts throughout a particular geographic area or time frame.

The choice of statistical method depends heavily on the nature of the collected data and the research questions being addressed. For instance, if the data is categorical (e.g., presence or absence of a particular feature), chi-square tests are appropriate. For continuous data (e.g., water depth, organism size), parametric tests (such as t-tests, ANOVA, or regression analysis) might be used. However, if the data fails to meet the assumptions of these parametric tests (e.g., normality, homogeneity of variance), non-parametric alternatives (such as Mann-Whitney U test, Kruskal-Wallis test, or Spearman correlation) should be employed. The selection process needs to be thoroughly documented to ensure the reproducibility and transparency of the investigation. Choosing appropriate tests based on the characteristics of the data is critical to obtaining accurate and reliable results. Inappropriate statistical methods can lead to erroneous conclusions and undermine the credibility of the investigation.

Furthermore, the limitations of each method must be explicitly acknowledged. Statistical significance does not necessarily imply practical significance. A statistically significant result might reflect a small, biologically inconsequential effect, especially with large sample sizes. Conversely, a non-significant result might not necessarily indicate the absence of an effect; it could simply reflect a lack of statistical power due to limited sample size or high variability in the data. Therefore, a comprehensive interpretation of statistical results requires careful consideration of effect sizes and the confidence intervals around the estimated effects. Presenting findings clearly and concisely, acknowledging limitations, and emphasizing the need for further investigation where appropriate contribute to the scientific credibility of cryptozoological studies.

The increasing availability of sophisticated statistical software packages, such as R or SPSS, has simplified the process of data analysis. However, it is crucial to possess a strong understanding of the underlying statistical principles to ensure the appropriate application of these tools. Misinterpreting or misusing these powerful tools can easily lead to erroneous conclusions. The focus should be on understanding the principles underlying statistical testing and hypothesis formulation, not just on the mechanics of using specific software. Understanding the underlying theory behind the statistical methods chosen allows for critical evaluation of the results obtained and a more informed discussion of the limitations and potential biases inherent in the study.

Beyond the application of formal statistical methods, a critical aspect of data analysis in cryptozoological investigations is the handling of uncertainty and ambiguity. Data collected in this field is often incomplete, potentially biased, and subject to considerable uncertainty. For example, eyewitness accounts of cryptids can be unreliable due to factors such as poor visibility, distance from the observed object, and the observer's emotional state. Photographic and video evidence can be blurry, obscured, or even misrepresented. Incorporating subjective evidence like eyewitness accounts requires careful consideration of potential biases and limitations, relying heavily on triangulation with other sources of evidence. Statistical modeling can be used to estimate the probability of a particular explanation given the available data and potential confounding factors, providing a way to quantify uncertainty. Bayesian methods, incorporating prior knowledge and incorporating updated probabilities based on new evidence, provide flexible methods to manage uncertainty.

In conclusion, statistical analysis is an indispensable component of any rigorous cryptozoological investigation. The methods described in this section, ranging from descriptive statistics to hypothesis testing and regression analysis, provide the tools for transforming raw data into meaningful insights. However, it is crucial to remember that the application of these statistical techniques must be guided by a clear understanding of their underlying principles, limitations, and the specific context of the investigation. Only through a combination of careful data collection, rigorous statistical analysis, and a critical appraisal of results can we move beyond mere speculation and develop a stronger, evidence-based understanding of the enigmatic creatures that may inhabit our oceans' depths. The successful application of statistical analysis in cryptozoological research not only strengthens the credibility of the field but also facilitates a deeper understanding of marine biodiversity and the vast unexplored realms of the ocean. This understanding, irrespective of the discovery of cryptids, benefits the broader scientific community and contributes to the conservation of marine ecosystems.

The preceding sections detailed the crucial role of statistical analysis in transforming raw data into meaningful insights within the context of marine cryptozoological investigations. However, the power of this analysis is only fully realized when the findings are effectively communicated. Data visualization is the bridge between complex statistical results and clear, understandable communication, crucial for convincing both the scientific community and the wider public of the validity of the research. Effective data visualization requires careful consideration of the type of data, the intended audience, and the specific message to be conveyed. Simply presenting raw numbers or complex statistical tables is rarely sufficient; compelling visuals are essential for engaging the reader and enhancing understanding.

One of the most common and versatile methods for visualizing data is through graphs. Different types of graphs are suited to different types of data. For instance, bar charts are ideal for comparing the frequencies of categorical variables. Consider a study investigating reports of a potential cryptid sighted near specific geographical locations. A bar chart could effectively represent the number of sightings at each location, instantly highlighting areas of higher sighting concentration. Similarly, line graphs are excellent for displaying trends over time. In analyzing historical sighting records, a line graph could illustrate the frequency of sightings over decades, revealing potential patterns of seasonal occurrence or long-term population trends. Pie charts, on the other hand, are effective for representing proportions or percentages within a whole, showing the relative contributions of different factors to a phenomenon, perhaps the distribution of reported cryptid characteristics (color, size, etc.) or the breakdown of different types of evidence collected.

Scatter plots are particularly valuable for exploring the relationships between two continuous variables. For example, in a study investigating the correlation between water temperature and the frequency of cryptid sightings, a scatter plot would visually depict the relationship between these two variables. The presence of a clear pattern or trend in the plot would suggest a potential correlation. This visualization method allows for a quick grasp of potential relationships that may need further investigation through statistical methods like regression analysis. By combining scatter plots with trend lines, the visual representation of the strength and direction of the relationship can be further emphasized and made readily understandable even to a non-expert audience.

More advanced plotting techniques offer further sophistication in data presentation. Heatmaps, for example, are particularly useful when dealing with large datasets showing patterns across multiple variables, such as spatial distributions of sightings, combined with environmental factors. A heatmap could visually represent the overlap of reported cryptid locations with specific oceanographic conditions like temperature, salinity, or depth, allowing for immediate identification of potential hotspots or habitat preferences. This visual approach allows the rapid exploration of complex multi-dimensional datasets and their relationships, revealing potential correlations not easily apparent in tabular data alone.

Beyond simple graphs, charts, and plots, maps are indispensable tools for visualizing geographical data. In cryptozoological research, maps play a vital role in showing the spatial distribution of sightings, providing a visual context for understanding the potential range and habitat of a cryptid. Different map types offer various levels of detail and precision. Simple point maps can show the location of individual sightings, while density maps can visually represent the concentration of sightings in different regions. Overlaying sighting data onto other geographical data layers, such as bathymetric maps (showing ocean depth), sea surface temperature charts, or habitat maps, creates rich visual representations that can reveal important ecological relationships. The power of geographical information systems (GIS) software greatly enhances the analysis and visualization of spatial data in cryptozoological research, enabling interactive exploration of multiple layers of data.

Data visualization is not limited to static images. Interactive visualizations, using software and web-based tools, allow researchers to explore their data dynamically. This interactive exploration fosters deeper understanding and insight, encouraging discovery of patterns and relationships that might be missed in static representations. Interactive dashboards, for example, allow users to filter data, zoom into specific regions of interest, and manipulate different variables to examine their impact on the visualized data. This interactive approach is particularly useful when dealing with large, complex datasets, enabling users to explore the data in a more meaningful and intuitive way than traditional static representations. Interactive three-dimensional visualizations can also provide powerful and immersive experiences, allowing for a deeper appreciation of spatial relationships and contextual information.

Beyond graphs, charts, and maps, a variety of other visual tools can greatly enhance data presentation. Infographics, for example, combine visual elements with concise text summaries to present complex information in an accessible and engaging manner. They are particularly useful for communicating findings to a broad audience beyond specialists. Flowcharts can visually represent the stages of an investigation, while diagrams can illustrate the relationships between different variables or hypotheses. Careful consideration of the visual design, color palettes, and overall layout is crucial in crafting an infographic that is not only visually appealing but also accurately conveys the essential data. Well-designed infographics can transform complex data into readily comprehensible and captivating visual stories.

The choice of visualization methods should always align with the specific data and the intended audience. While sophisticated visualizations can be powerful, simplicity and clarity should always be prioritized. Avoid cluttering visualizations with too much information; instead, focus on highlighting the key findings and insights. Moreover, it's crucial to maintain ethical standards in data representation. Avoid manipulating visualizations to misrepresent findings or create a misleading impression. Transparency is paramount; the methods used for data collection, analysis, and visualization should be clearly documented and readily available for scrutiny. This transparency builds trust and allows others to reproduce the findings and verify their validity.

Furthermore, effective data presentation necessitates clear and concise labeling of all visual elements. Axes should be clearly labeled with units, legends should be comprehensive, and any relevant contextual information should be provided. The choice of colors, fonts, and visual style should be consistent and adhere to established guidelines for accessibility, ensuring that the visualization is easily understandable to a wide audience, including those with visual impairments. Providing a clear and concise narrative accompanying the visuals is crucial; the visuals should not stand alone but should be integrated into a larger narrative that contextualizes the findings and guides the reader through the interpretation of the data.

In conclusion, data visualization is not merely an aesthetic exercise; it is an integral part of the scientific process, transforming raw data into compelling narratives. By thoughtfully choosing and implementing appropriate visualization methods, researchers can effectively communicate their findings to both specialist and lay audiences, promoting greater understanding and appreciation of the complexities of marine cryptozoology. The power of visualization lies in its ability to bridge the gap between abstract data and intuitive understanding, transforming complex statistical results into readily accessible and compelling stories. Through a strategic combination of clear visual representations and concise narratives, the credibility and impact of cryptozoological research can be significantly enhanced. The ultimate goal is to communicate the findings in a way that is both persuasive and accurate, allowing readers to engage with the data and draw their own informed conclusions. The rigorous application of visualization techniques ultimately contributes to the advancement of the field and enhances its scientific credibility.

Interpreting the results of a cryptozoological investigation requires a nuanced approach, blending rigorous scientific methodology with an understanding of the inherent uncertainties involved. While statistical analyses provide valuable insights, they are not foolproof. The process of drawing conclusions necessitates careful consideration of several crucial factors, including the limitations of the data, the potential for biases, and the context within which the findings are situated. Simply because a statistical analysis reveals a correlation doesn't automatically imply causation. For instance, a strong correlation between reported sightings of a supposed cryptid and specific oceanographic conditions (e.g., temperature, salinity, currents) might suggest a habitat preference. However, it does not definitively prove that the cryptid exists or that the observed correlation isn't the result of some other, yet-unidentified factor.

The quality of the data itself is paramount. Eyewitness accounts, often the cornerstone of cryptozoological investigations, are susceptible to various biases, including misidentification, exaggeration, and memory distortion. Photographic or video evidence, while seemingly more objective, can also be ambiguous or misleading. The resolution, lighting conditions, and angle of capture can significantly affect the interpretation of the image. Furthermore, the possibility of image manipulation or misrepresentation needs careful consideration. The reliability of the source of the evidence—whether a seasoned observer or an unreliable witness—must be critically assessed. Establishing a robust chain of custody for physical evidence is crucial to maintain its integrity and ensure its admissibility in any future evaluation.

Environmental factors can also significantly influence the interpretation of findings. For example, unusual weather patterns, unusual ocean currents, or bioluminescent displays could all contribute to misinterpretations of what might be perceived as a cryptid sighting. Thorough environmental data collection, including water temperature, salinity, currents, and biological activity, is essential for contextualizing the observed phenomena and ruling out alternative explanations. This might involve consulting with oceanographers, meteorologists, or other relevant specialists to gain a comprehensive understanding of the environmental context. It is particularly crucial to understand the natural variability of the environment and account for natural fluctuations in ecological parameters that might mimic patterns associated with cryptid activity.

Statistical significance does not always equate to biological significance. A statistically significant result might indicate a real effect, but the magnitude of that effect could be negligible in a real-world context. For instance, a statistically significant increase in the frequency of sightings in a particular area might be entirely explained by an increase in human activity or improved observation techniques, rendering it insignificant from a cryptozoological perspective. Therefore, a comprehensive assessment of both statistical and biological significance is essential before drawing firm conclusions.

The inherent limitations of the study design should be openly acknowledged. The scope of the investigation, the sample size, and the methodology employed all impose limitations on the generalizability of the findings. For example, a study limited to a specific geographic area or time period cannot reliably extrapolate its conclusions to other regions or timeframes. Likewise, a small sample size can limit the statistical power of the analysis, making it difficult to detect subtle effects or patterns. Transparency about these limitations is critical to maintain the integrity of the research and prevent overreaching conclusions. Such transparency allows the research community to assess the validity of the findings and to identify avenues for future investigation.

Furthermore, the possibility of alternative explanations should always be considered. Before attributing an observation to an unknown cryptid, all known explanations, however improbable they may seem, must be rigorously investigated and ruled out. This might involve consulting with experts in related fields, such as marine biology, oceanography, or geology, to assess the plausibility of alternative explanations. A thorough literature review is also crucial, ensuring that similar phenomena have been appropriately investigated and discounted.

The communication of research findings is another critical aspect of the investigative process. The findings should be presented in a clear, concise, and unbiased manner, avoiding sensationalism or speculation. The methodology employed, the limitations of the study, and the potential for alternative explanations should be transparently communicated. Visual aids, such as graphs, maps, and charts, can be instrumental in conveying complex information effectively. This enhances accessibility and comprehension for a wider audience. Furthermore, the language used should be accessible to both specialists and the general public, avoiding jargon and technical terminology that may hinder understanding.

Disseminating research findings requires careful consideration of the intended audience. Scientific publications are valuable for rigorous peer review and validation within the scientific community. Conversely, public presentations or popular science articles can educate a broader audience and cultivate greater interest and support for future research. Choosing the appropriate platform and communication style to engage each particular audience is essential. The goal is not merely to disseminate information but also to foster dialogue, collaboration, and further critical analysis of the findings.

In conclusion, interpreting results and drawing conclusions from cryptozoological investigations demand a rigorous and multifaceted approach. Combining careful data analysis, cautious interpretation, a willingness to consider alternative explanations, and transparent communication of findings are all essential for advancing knowledge in this field. While the pursuit of sea cryptids involves inherent uncertainties, a commitment to scientific rigor ensures that investigations remain grounded in evidence and contribute meaningfully to our understanding of the marine environment. By acknowledging limitations, transparently reporting findings, and continuously refining methodologies, we can strengthen the credibility of cryptozoological research and make progress in unraveling the mysteries of the deep. This careful and balanced approach ensures that investigations remain rigorous, ethical, and productive, even when dealing with the inherently challenging subjects of cryptozoology.

The pursuit of cryptids, particularly those inhabiting the vast and largely unexplored ocean depths, presents unique challenges to data analysis and interpretation. The inherent uncertainties associated with underwater observation, the often-limited and fragmentary nature of evidence, and the difficulties in controlling variables all contribute to the complexities involved. Therefore, a rigorous approach to error analysis and uncertainty quantification is essential to ensure the reliability and validity of any findings.

One of the primary sources of error in cryptozoological investigations lies in the data itself. Eyewitness accounts, often the foundation of such investigations, are notoriously subjective. Factors such as individual perception, memory biases, environmental conditions, and the inherent difficulty in identifying unfamiliar creatures contribute to a high degree of uncertainty. The reliability of witnesses needs careful assessment. Experience with marine life, training in observation, and any potential biases must all be considered. Triangulating accounts from multiple independent sources, if possible, can help improve reliability but doesn't eliminate individual biases completely. Similarly, corroborating eyewitness accounts with physical evidence, such as photographs or videos, can help strengthen the case, although even these sources are not without their limitations.

Photographic and video evidence, while often considered more objective, are also subject to significant uncertainties. The quality of the image or video is crucial. Poor resolution, inadequate lighting, and challenging underwater conditions can all obscure details, making identification extremely difficult. The angle of capture can also significantly influence the interpretation of the image, leading to misinterpretations of size, shape, and features. Furthermore, the possibility of image manipulation or misrepresentation must always be considered, requiring careful scrutiny of metadata and origin. Advanced image analysis techniques, such as photogrammetry, can help extract additional information but should be applied cautiously and with a thorough understanding of their limitations. Sophisticated image processing tools can enhance the data but cannot create information that isn't already present in the original image.

Acoustic data, such as sonar recordings, provide another avenue for investigating cryptids. However, sonar readings are susceptible to various types of interference and artifacts, including reverberation, multipath propagation, and environmental noise. These artifacts can mimic the presence of objects that are not actually there, leading to false positives. Careful calibration and signal processing are essential to minimize these errors. Additionally, the interpretation of sonar signals requires expertise in sonar technology and an understanding of the acoustic properties of the marine environment. Distinguishing a cryptid's acoustic signature from that of known marine life or natural phenomena is challenging and often requires comparative analysis with known sound profiles.

Environmental factors significantly impact the reliability of data collected in cryptozoological investigations. Variations in water temperature, salinity, currents, and light levels can influence the behavior and visibility of marine organisms, potentially affecting the detection and identification of cryptids. Therefore, environmental data collection is paramount. Factors such as water depth, seabed topography, and the presence of other marine life should be considered and carefully documented. Comparing data with historical records of environmental conditions in the same area can help to establish a baseline and identify unusual deviations, thus allowing a researcher to rule out anomalous environmental events as explanations for the observed phenomenon. Understanding how environmental variables might influence both the behavior of the cryptid and the probability of its detection is also critical.

Statistical analyses play a crucial role in analyzing cryptozoological data, but their interpretation needs caution. It's crucial to recognize that statistical significance does not always equate to biological or practical significance. A statistically significant result merely indicates that the observed pattern is unlikely to have occurred by random chance. However, it doesn't necessarily imply a causative relationship or that the effect is large enough to be biologically meaningful. For example, a statistically significant increase in the frequency of cryptid sightings in a particular region could be entirely attributable to an increased number of observers in that area, improved observation technologies, or heightened media attention rather than a real increase in the cryptid population.

Uncertainty quantification involves assessing the range of possible values for a particular parameter, acknowledging the limitations of the data and the methodologies used. This often involves the calculation of confidence intervals, which provide a range of values within which the true value of a parameter is likely to lie, given a specified level of confidence. Similarly, the use of Bayesian statistical methods allows for incorporating prior knowledge and uncertainty into the analysis, leading to a more nuanced understanding of the findings. Robust uncertainty quantification helps determine the significance of the results, and highlights the areas requiring further investigation to reduce uncertainty.

Addressing potential sources of error is crucial in enhancing the reliability of cryptozoological research. This involves carefully considering all possible explanations, including those seemingly improbable. Consulting with experts in relevant fields, such as marine biology, oceanography, geology, and acoustics, can help rule out alternative hypotheses and identify potential biases. A thorough literature review is essential, ensuring that similar phenomena have been investigated appropriately and that any potential biases have been accounted for.

The importance of transparency in reporting results cannot be overstated. Researchers must openly acknowledge the limitations of their study, including the sample size, the methodology employed, and any potential sources of bias. Presenting data clearly and avoiding sensationalism or speculation is paramount. The use of visual aids, such as maps, graphs, and charts, can significantly enhance the understanding and accessibility of research findings, making them more comprehensible to a wider audience. Utilizing plain language that is understandable to both specialists and the general public avoids the potential for misinterpretations and fosters better communication.

In conclusion, error analysis and uncertainty quantification are essential components of rigorous cryptozoological investigations. A meticulous approach to data collection, analysis, and interpretation, coupled with an acknowledgment of limitations and a willingness to consider alternative explanations, is vital for ensuring the reliability and validity of research findings. By embracing transparency, employing robust statistical methods, and actively seeking to reduce uncertainties, we can improve the credibility and scientific rigor of cryptozoological research, furthering our understanding of the ocean's mysteries and potentially revealing some of its long-hidden secrets. This commitment to rigorous methodology allows researchers to contribute meaningfully to our understanding of the marine world, regardless of the ultimate outcome of their quest for elusive cryptids. The goal remains the advancement of knowledge, a goal achievable even with the inherent uncertainties involved in this unique field of study.

The application of rigorous data analysis and interpretation is crucial to separating credible evidence from anecdotal accounts and misinterpretations in cryptozoological investigations. This requires a multi-faceted approach, encompassing various data types and analytical techniques. Let's examine several case studies that illustrate this process.

One notable example involves the analysis of purported "giant squid" sightings. Historically, these accounts have ranged from fishermen's tales of colossal tentacles thrashing about to blurry photographs and damaged fishing nets. Analyzing these diverse reports requires a nuanced approach. We begin by geographically mapping the locations of reported sightings. Clustering of sightings in particular areas might indicate a higher probability of genuine encounters, or alternatively, a regional bias in reporting (perhaps influenced by local legends or media attention).

Next, we analyze the descriptions of the creature itself. Are there consistent features across accounts, or is there too much variability to suggest a single species? Detailed comparisons of body proportions, coloration, and behavioral descriptions from different eyewitness accounts can reveal patterns or inconsistencies. Furthermore, the time of year of each sighting may reveal seasonal migration patterns or breeding behaviors that could corroborate the existence of a specific species. Analysis of historical accounts, including those from the 19th and early 20th centuries, often contains valuable information, although one must critically assess the accuracy and potential biases present in older records.

Further investigation might involve examining physical evidence. Photographs and videos, even if low-resolution, can be analyzed using image enhancement techniques to identify potentially identifying features. For instance, advanced digital image processing can sharpen blurry images, improving visibility of key features. In cases where physical remains exist, such as damaged fishing equipment or tissue samples, genetic analysis can offer valuable insights. DNA sequencing allows for comparison against known species, helping to determine whether the specimen represents a known organism or an entirely new one. However, it's crucial to acknowledge the limitations of DNA analysis in this context. Degraded DNA might yield inconclusive results, and contamination must be carefully controlled to avoid false positives.

Another compelling case study involves the investigation of unusual acoustic signals recorded in the deep ocean. Sonar data often presents ambiguous results. Distinguishing between natural phenomena, such as underwater landslides or whale calls, and potential cryptid sounds requires careful spectral analysis. The frequency, amplitude, and temporal patterns of the acoustic signals are compared against a database of known marine sounds. The use of machine learning algorithms can help automate this comparison and identify potential anomalies that warrant further investigation.

However, it's crucial to acknowledge that novel sounds can be simply new sounds from previously unknown animals. Equating the presence of unknown sounds directly to an unknown creature requires considerable caution and should be rigorously justified. The location and timing of acoustic signals are also important factors to consider. Repeated signals emanating from a specific area over time might suggest the presence of a persistent source, whereas sporadic signals could be attributed to unusual events or random occurrences. Correlation with other data types, such as eyewitness accounts or physical evidence from the same region, is essential to strengthen the case.

The analysis of supposed "sea serpent" sightings also provides valuable lessons. Many sightings can be attributed to misidentifications of known marine animals, such as basking sharks, oarfish, or giant squid. By comparing descriptions and eyewitness testimony with the morphological characteristics and behavioral patterns of known species, we can often eliminate seemingly credible reports. The development of a comprehensive database of marine life, including their size ranges, habitats, and behavioral traits, facilitates this process. Computer simulations that model the appearance of marine animals from various angles and at different distances can also help to resolve discrepancies and better match witnessed sightings to likely culprits. The reliability of eyewitness testimony needs careful evaluation. Factors such as distance from the subject, environmental conditions, and the witness's level of marine biology expertise all contribute to the accuracy of a description. The use of statistical tools such as Bayesian networks can help incorporate this uncertainty into an overall assessment of the reliability of the account. These networks allow researchers to update their assessment of probabilities as additional evidence becomes available.

Let us also consider cases involving underwater drone exploration. Advanced underwater drones equipped with high-resolution cameras and sonar can gather detailed data about the ocean's depths. Video footage from these drones can be analyzed to identify potential cryptids or unusual phenomena. However, the interpretation of this data requires expertise in marine biology and underwater technology. It is paramount to rule out alternative explanations for unusual objects observed in the videos. Technical artifacts, such as light refraction or camera malfunction, must be considered and eliminated. Similarly, debris, geological formations, or aggregations of marine life could be mistakenly identified as cryptids.

In each of these case studies, the rigorous application of statistical analysis is paramount. Simple descriptive statistics, such as frequencies and proportions, provide a basic understanding of the data. More sophisticated techniques, such as regression analysis, can explore the relationships between different variables. For instance, we could examine the correlation between water temperature and the frequency of cryptid sightings. However, correlation doesn't imply causation. Careful consideration of confounding variables is vital to avoid drawing spurious conclusions. The use of advanced statistical modeling, such as hierarchical Bayesian models, can account for uncertainty and incorporate prior knowledge into the analysis. Moreover, rigorous hypothesis testing ensures that any conclusions drawn are supported by statistical evidence and are not simply the product of chance. A p-value, although heavily scrutinized and debated within the scientific community, provides a quantifiable metric to evaluate the plausibility of random chance as an explanation of the observed data.

In conclusion, the analysis and interpretation of data in cryptozoological investigations require a multi-faceted approach, combining various data types, analytical techniques, and a critical assessment of potential biases and limitations. The case studies examined demonstrate that rigorous methodology, transparency in reporting, and a commitment to scientific rigor are essential to advancing our understanding of the ocean's mysteries, whatever the final conclusion may be. The advancement of knowledge, not merely the confirmation of preconceived notions, remains the ultimate goal. Only through this dedicated approach can we hope to separate credible hypotheses from speculation and further scientific knowledge in the field of marine biology, including the exciting possibility of the discovery of new species.

9

Presenting Your Findings

The culmination of any cryptozoological investigation, regardless of whether it involves the elusive giant squid or the enigmatic sea serpent, lies in the presentation of findings. This isn't merely a matter of compiling data; it's about crafting a compelling, scientifically rigorous narrative that persuades the reader of the validity of your methods and the reliability of your conclusions. This necessitates a well-structured scientific report, a document that transcends mere storytelling and adheres to established academic standards. This report should serve as a testament to the meticulous nature of your investigation, conveying not just your findings, but also the process by which you arrived at them.

The structure of a scientific report, although adaptable to specific contexts, generally follows a consistent framework. It begins with an abstract, a concise summary of the entire report, encapsulating the research question, methods, key findings, and conclusions. This serves as a roadmap for the reader, providing a bird's-eye view of your work before delving into the details. Think of it as a compelling elevator pitch for your research.

The introduction, expanding on the abstract, lays the groundwork for your investigation. It should clearly articulate the research question—the central puzzle your investigation aims to solve. This might be framed as a specific hypothesis, a testable statement predicting a particular outcome, or a more open-ended exploration of a phenomenon. The introduction also provides crucial background information, contextualizing your research within the broader field of marine biology and cryptozoology. This section should review relevant literature, demonstrating your awareness of existing knowledge and identifying the gaps your research aims to fill. Here, you'll cite established scientific sources and demonstrate a deep understanding of the subject matter, showcasing the foundation upon which your investigation is built.

The methodology section is the heart of a scientifically rigorous report. It meticulously details the techniques and procedures you employed during your investigation. This should leave no room for ambiguity. For example, if you used sonar, specify the type of sonar used, its frequency range, the sampling rate, and any signal processing techniques employed. If you analyzed visual data, describe the image enhancement techniques utilized, the software employed, and any limitations inherent in the methods. If eyewitness accounts formed part of your analysis, explain how you assessed their reliability and accounted for potential biases. This meticulous approach ensures reproducibility; another researcher should be able to replicate your study using your precise methodology. This is the cornerstone of scientific credibility.

The results section focuses solely on the presentation of your data, avoiding any interpretation or speculation. This section employs graphs, charts, tables, and other visual aids to present your data in a clear and concise manner. Each figure or table should be carefully labeled and captioned, and its relevance to the overall findings should be clearly stated. For instance, if you mapped sightings of an unusual marine organism, the map should clearly show the geographic distribution of sightings, including dates and any other relevant information. Likewise, if you analyzed sonar data, the results should be presented in a format that allows for easy comparison with known marine sounds. Raw data should be included, either within the report itself or in an appendix, to ensure transparency and allow for independent verification. The clarity and organization of this section are critical to ensuring readers can fully grasp your findings.

The discussion section is where you interpret your findings and relate them back to your research question. This section goes beyond the mere presentation of data, offering a nuanced interpretation of its implications. Here, you discuss the significance of your findings, addressing their strengths and limitations. This may involve comparing your results with predictions from established models or theories. Furthermore, this section addresses any unexpected findings or anomalies, acknowledging their potential significance while emphasizing the need for further investigation. Consider alternative explanations for your results, exploring potential biases and limitations in your methodology. You must acknowledge the possibility that your conclusions are not definitive, especially in a field as challenging as cryptozoology.

The conclusion summarizes your key findings and their implications, returning to the central research question posed in the introduction. Does your research support a specific hypothesis? Does it suggest the existence of a cryptid, or does it provide evidence contrary to such a claim? This section reiterates the main takeaways of your report in a concise and accessible manner, drawing a clear and decisive conclusion. This concise summary reinforces the key messages of the report.

Finally, the references section lists all the sources you cited throughout your report, following a consistent citation style (such as APA or MLA). This section demonstrates the breadth of your research and acknowledges the contributions of others in the field. It's crucial to ensure complete and accurate referencing to avoid plagiarism and uphold the ethical standards of scientific research.

Beyond the formal structure, the language used in your scientific report should be precise and objective. Avoid colloquialisms, jargon, and subjective language. Instead, use clear, concise, and unambiguous language to convey your information accurately. Graphs, charts, and other visual aids should enhance understanding and not complicate it. Your ultimate aim is to present your findings in a manner that is both scientifically rigorous and easily comprehensible to a broader audience. Remember that the goal is to advance our understanding of the ocean's mysteries, not solely to promote a specific conclusion. Maintaining scientific rigor and integrity are essential in producing a trustworthy and compelling scientific report, regardless of the nature of your findings. The pursuit of knowledge, even in the fascinating but sometimes ambiguous realm of cryptozoology, demands the highest standards of scientific methodology and reporting. Only through this approach can we hope to distinguish sound scientific conclusions from unfounded speculation.

Having meticulously documented your findings in a comprehensive scientific report, the next crucial step involves presenting your research to an audience. This presentation, whether delivered at a scientific conference, a public lecture, or even a smaller, more informal gathering, is your opportunity to share your work, engage with others in the field, and potentially spark further investigation.

Preparing a compelling and effective presentation requires careful planning and consideration of several key elements.

The first critical decision involves selecting the appropriate format for your presentation. The length and nature of your presentation will greatly influence the most effective approach. A short presentation, perhaps for a conference, might lend itself to a concise overview, focusing on the key findings and their implications. This might involve a rapid-fire succession of visually engaging slides, each summarizing a key aspect of your research. Conversely, a longer presentation, such as a public lecture, affords greater flexibility. You might incorporate more narrative elements, weaving together a compelling story that draws the audience into the world of your investigation. Consider using a narrative structure to enhance engagement—think of it as a journey of discovery, leading your audience step-by-step through your investigative process.

Regardless of length, a clear and logical structure is paramount. A well-organized presentation should follow a narrative arc, moving smoothly from the introduction to the conclusion. Start by clearly stating your research question, providing essential background information, and then detailing your methodology. This establishes the context and credibility of your work. Next, present your key findings, using clear and concise language and avoiding jargon. Support your findings with compelling visuals, and then discuss the implications of your work. Conclude with a strong summary and a call to action, inspiring further research or discussion. This structure ensures a cohesive and impactful presentation.

Visual aids are integral to an effective presentation. Well-chosen visuals can greatly enhance understanding and engagement, making even complex information accessible to a broader audience. However, it's crucial to avoid overwhelming the audience with too much visual information. Keep slides clean and uncluttered, using high-quality images and graphics. Consider using maps to illustrate geographic locations, graphs and charts to visualize data, and photographs or video footage to showcase your fieldwork. Each visual should serve a specific purpose and directly support your narrative. For example, if presenting sonar data, clearly labeled spectrograms can significantly enhance comprehension compared to a complex table of numerical data. Similarly, showcasing high-resolution underwater images or video taken by remotely operated vehicles (ROVs) can be far more compelling than simply describing the findings verbally. Remember to cite the source of all images and videos to maintain academic integrity.

The selection of visual aids is also crucial. Avoid cluttered slides filled with dense text. Instead, opt for impactful imagery and minimal text. Let the visuals tell the story, and use text only to highlight key points or provide concise labels. Consider incorporating animations or transitions subtly to maintain audience engagement, but avoid excessive animation that can be distracting. The quality of your images and graphics is crucial. Ensure that all visuals are high-resolution and professionally presented. Poor-quality visuals can detract from your presentation and undermine its credibility. The overall aesthetic consistency of your slides also greatly impacts the audience's perception of your research.

Your delivery style is equally important as the content itself. A confident and engaging delivery can transform a potentially dry presentation into a captivating experience. Practice your presentation thoroughly beforehand, ensuring a smooth and natural flow. Speak clearly and maintain eye contact with your audience. Vary your tone and pace to maintain interest and avoid monotony. Use pauses effectively to emphasize key points and allow your audience time to process the information. Enthusiasm is contagious; let your passion for your subject shine through, and engage your audience with your energy and commitment to the subject matter. Consider incorporating personal anecdotes or stories to make the presentation relatable and memorable. This creates a connection with your audience and makes the information more engaging. It's perfectly acceptable to add a touch of your personality while maintaining scientific rigor. The goal is to create a strong connection with the audience, making your research not just informative, but also memorable.

Depending on your audience, you may need to adjust your delivery and level of technical detail. A presentation for a scientific conference will likely require a higher level of technical expertise than a presentation for the general public. For specialized audiences, you can confidently use technical jargon and delve into intricate details. However, when addressing a more general audience, it is vital to simplify complex concepts, avoiding jargon and using clear, concise language. Tailor your delivery to your audience and their level of understanding.

Anticipate questions from the audience and prepare thoughtful and concise answers. It is vital to acknowledge any limitations in your research and to remain open to constructive criticism. Be prepared to discuss alternative interpretations of your data and to engage in respectful and productive dialogue. The Q&A session is an integral part of the presentation, providing a valuable opportunity to engage with your audience and to further refine your understanding of the subject matter. It also serves as an opportunity to expand on points that may have not been explicitly covered.

Consider conducting a test run before the actual presentation. Present your work to a small, trusted group of colleagues or friends. This allows you to receive constructive feedback on the clarity, pacing, and overall effectiveness of your presentation. It provides a valuable opportunity to identify any areas that need improvement before the final presentation. This feedback is crucial in ensuring a smooth and well-received presentation. You can use their feedback to refine your visuals, refine your narrative, and enhance your delivery style. This trial run will significantly improve your confidence and prepare you for the actual presentation.

After the presentation, reflect on your performance. What aspects of the presentation were particularly effective? What areas could be improved? Analyzing your successes and areas for improvement will help you refine your presentation skills over time. Take notes and incorporate the constructive criticism into your subsequent presentations to further enhance the quality and engagement. This reflective practice is critical for improving your presentation skills and ensuring that you effectively communicate your research findings.

Presenting your research is an important step in the process of disseminating knowledge and potentially contributing to a broader understanding of the world around us, even within the intriguing and sometimes controversial realm of cryptozoology. By meticulously planning, crafting a compelling narrative, and employing effective communication techniques, you can create a lasting impression on your audience and inspire future investigations into the mysteries that lie beneath the waves. Remember, the presentation is not just about conveying facts and figures; it's about sparking interest, inspiring curiosity, and ultimately, advancing our collective understanding of the enigmatic world of sea cryptids.

The culmination of months, perhaps years, of painstaking research doesn't end with a neatly bound scientific report gathering dust on a shelf. The true value of your investigation lies in its dissemination – sharing your findings with the wider scientific community and, importantly, the public. This process is multifaceted, requiring strategic approaches tailored to different audiences and platforms. Simply put, the most rigorously conducted research is useless unless it reaches those who can benefit from it, furthering the understanding of marine biodiversity and our planet's hidden wonders, even those that reside in the realm of cryptozoology.

One of the most traditional, and arguably still the most prestigious, avenues for disseminating scientific findings is through peer-reviewed publications. Submitting your work to a reputable scientific journal, such as those specializing in marine biology, zoology, or even cryptozoology (if such a dedicated journal exists), subjects your research to rigorous scrutiny. The peer-review process, though demanding, ensures the quality and validity of your work, lending it credibility within the scientific community. The process typically involves submitting a detailed manuscript, including methodology, data analysis, results, and discussion, to the journal's editors. These editors then send the manuscript to several experts in the relevant field for independent evaluation. Their assessments, often including constructive criticism, are crucial for refining your work before publication. A well-written and thoroughly researched manuscript, adhering to the journal's strict guidelines, significantly increases the likelihood of acceptance.

Remember to clearly and concisely articulate your findings, support your claims with robust evidence, and acknowledge any limitations of your study. Publication in a reputable journal elevates your research beyond a local context, making it accessible to a global audience of scientists and researchers. The potential for collaborative opportunities and further investigation arising from such publication is considerable.

Beyond peer-reviewed publications, scientific conferences and symposia provide a dynamic platform for presenting your work. These events typically feature presentations, poster sessions, and networking opportunities, allowing for immediate feedback and collaborative discussions. Preparing an effective conference presentation requires a concise and engaging summary of your findings, suitable for a time-constrained format. Visual aids, such as slides with high-quality images and graphs, are essential to conveying complex information clearly and engagingly. Practicing your presentation beforehand is crucial, ensuring a confident and smooth delivery. The Q&A session following your presentation provides a valuable opportunity for interaction with experts in the field, offering valuable insights and potential directions for future research. Networking with fellow researchers during the conference can also lead to collaborations, shared expertise, and potential future research ventures. Even seemingly tangential connections can lead to important breakthroughs, demonstrating the interdisciplinary nature of scientific discovery.

Public engagement, though often overlooked, holds significant value in disseminating research findings beyond the scientific community. Engaging the public fosters broader awareness, generates support for conservation efforts, and ultimately contributes to a more informed and environmentally conscious citizenry. This can take many forms, from public lectures and talks at museums and community centers, to contributions to popular science magazines and articles, or participation in documentaries and television programs. When presenting your findings to a non-scientific audience, clarity and simplicity are paramount. Avoiding technical jargon and using illustrative examples are essential for enhancing understanding and engagement. Visually rich presentations, incorporating compelling imagery and videos, can greatly improve audience reception. Maintaining an engaging and approachable tone, sharing personal anecdotes, and incorporating elements of storytelling can enhance the impact and memorability of your presentation. Furthermore, participation in educational outreach programs, targeting students and young people, can inspire future generations of scientists and conservationists.

The use of digital platforms and online tools represents a significant advancement in disseminating scientific information. Websites, blogs, and social media channels provide a vast reach for your findings, transcending geographical barriers and reaching a diverse audience. Creating a dedicated website for your research project allows for detailed dissemination of your findings, including detailed reports, photographic evidence, and video footage from your expeditions. The use of blogs and social media, such as Twitter or Instagram, enables you to share concise updates, progress reports, and engaging visuals with a wider audience in a timely manner. However, the use of digital platforms requires careful consideration of ethical implications, including accuracy, transparency, and responsible engagement with online communities.

Maintaining a high standard of accuracy and transparency in your communications is critical to building trust and maintaining the credibility of your work. Actively engaging with comments and questions from online audiences fosters a two-way dialogue, leading to deeper understanding and potentially invaluable feedback. Furthermore, it's essential to be aware of misinformation and address any potential inaccuracies or misleading interpretations of your work proactively. The digital realm offers a remarkable opportunity to engage with a vast and diverse audience, but it requires responsible and ethical stewardship.

Finally, collaboration is paramount in disseminating research findings effectively. Networking with other researchers, conservation organizations, and government agencies enhances the reach and impact of your work. Collaborating with scientific institutions can provide access to wider resources and expertise, further refining your investigation and its dissemination. Partnering with conservation organizations can strengthen the practical application of your findings and promote conservation initiatives. Working with government agencies can ensure your research informs policy decisions and environmental regulations. Such collaborations increase the potential influence of your research, ensuring its relevance extends beyond the academic realm.

In conclusion, the dissemination of your findings, even in the sometimes controversial field of cryptozoology, is not merely a final step, but a crucial ongoing process. By employing a range of strategies, from formal scientific publications to public engagement initiatives and leveraging digital platforms, you can ensure your research reaches its intended audience, further advancing our understanding of the ocean's mysteries and promoting a responsible and informed approach to marine conservation. This multifaceted approach ensures the long-term impact of your investigation, solidifying its place within the scientific record and inspiring future generations to explore the unexplored depths of our planet's oceans. Remember, your research isn't simply data; it's a narrative waiting to be shared, a story of exploration, discovery, and the relentless pursuit of understanding the enigmatic world beneath the waves. The effective dissemination of this narrative is the key to its lasting impact.

The dissemination of research, particularly in a field as inherently speculative as cryptozoology, inevitably invites scrutiny. While rigorous methodology and transparent data presentation are crucial for building credibility, anticipating and addressing criticism is equally vital. The scientific community, renowned for its skepticism, often demands a high level of evidence before accepting unconventional findings. The public, meanwhile, may approach the subject with a mix of fascination and disbelief, shaped by popular culture portrayals and personal biases. Therefore, a successful investigation requires not just meticulous data collection and analysis, but also a well-developed strategy for responding to skepticism and criticism effectively.

One of the most common criticisms leveled against cryptozoological research concerns the nature of evidence. Eyewitness accounts, often the foundation of many claims, are notoriously unreliable, prone to misinterpretation, exaggeration, or even outright fabrication. Photographic and video evidence, while seemingly more objective, can be easily manipulated or misinterpreted due to poor resolution, lighting conditions, or the inherent difficulty of capturing elusive creatures in their natural environment. Addressing these concerns requires acknowledging the inherent limitations of each type of evidence and employing robust methods to validate findings. For instance, if relying on eyewitness testimony, triangulating accounts from multiple independent sources can help assess consistency and reliability. Similarly, analyzing photographic and video evidence using advanced image processing techniques and expert consultation can help verify authenticity and identify potential misinterpretations. It is crucial to present the evidence clearly, acknowledging limitations and emphasizing the rigorous analysis employed in evaluating it. Transparency is key; openly discussing potential biases and uncertainties demonstrates a commitment to scientific integrity and reduces the impact of skepticism.

Another frequent area of criticism revolves around research methodology. Critics may question the sampling methods, data analysis techniques, or the overall design of the study, arguing that the approach wasn't sufficiently rigorous or controlled to draw conclusive conclusions. Addressing such criticisms requires a thorough explanation of the methodology employed, justifying every step in the research process. This includes detailing the rationale behind the chosen methods, outlining any limitations, and explaining how these limitations were addressed or mitigated. Clearly articulating the statistical methods used for data analysis and presenting the results transparently, including any statistical uncertainties, is critical. Furthermore, it is beneficial to engage with the existing scientific literature on the subject, demonstrating familiarity with established methodologies and explaining any deviations from standard practices. This shows that the research was informed by existing knowledge and that any unique approaches were carefully considered and justified.

Specific criticisms will vary depending on the nature of the findings. For instance, if the research proposes a new species, it will be subject to intense scrutiny regarding taxonomic classification, genetic analysis, and morphological comparisons with known species. The lack of a physical specimen for DNA analysis might be criticized as severely limiting the validity of the claim. In this case, a detailed response should address these points directly, highlighting the evidence supporting the proposed classification, explaining the limitations posed by the lack of a physical specimen, and proposing alternative lines of inquiry that could further support the claim. Perhaps environmental DNA analysis or the development of advanced imaging techniques could provide supplemental evidence.

Responding to public skepticism requires a different approach. Public understanding of science, and particularly cryptozoology, is frequently influenced by popular culture, leading to unrealistic expectations and a predisposition to interpret evidence through the lens of established narratives (like the Loch Ness Monster legend). Addressing this involves using clear, concise language, avoiding technical jargon, and employing illustrative examples to make the research accessible to a wider audience. Visual aids, such as photographs, maps, and videos, can be highly effective in conveying complex information and fostering engagement. Sharing personal anecdotes about the fieldwork and the investigative process can make the research more relatable and engaging. Active engagement with public comments and questions, both online and offline, is crucial in addressing misconceptions and building trust. This can involve participating in public forums, giving lectures, or responding to inquiries from media outlets. Maintaining an open dialogue and acknowledging legitimate concerns demonstrates a commitment to transparency and credibility.

Furthermore, it's essential to acknowledge the limitations of the research. No scientific investigation is perfect, and admitting limitations upfront enhances credibility rather than diminishes it. Highlighting any uncertainties or areas where further investigation is needed shows a commitment to scientific rigor and self-reflection. This is especially important in cryptozoology, where definitive proof is often elusive. Emphasizing the importance of future research, proposing specific experiments or observational studies, and inviting collaboration from other researchers can further strengthen the impact and credibility of the work.

Finally, the ethical considerations involved in cryptozoological research are crucial. Respect for the environment and wildlife is paramount. If investigating a potential cryptid in a sensitive ecosystem, ensuring that the research minimizes any environmental impact is crucial. This may include obtaining necessary permits and adhering to strict guidelines for wildlife observation and data collection. Transparency in the research process, ensuring all data is properly documented and accessible for review, is a critical aspect of ethical research. Acknowledging potential biases and limitations of the study, as well as the impact of media attention on the subject and its potential habitat, contributes to the responsible conduct of research.

In conclusion, responding to criticism and skepticism requires a proactive and multifaceted approach. It's not about dismissing dissenting voices but engaging with them constructively, addressing concerns with evidence-based arguments and a commitment to transparency. By employing rigorous methodology, presenting findings clearly and concisely, acknowledging limitations, engaging with the public, and adhering to ethical standards, researchers can foster a deeper understanding of their work and its potential contributions to scientific knowledge, even within the often controversial realm of cryptozoology. The successful navigation of this process enhances the credibility of the findings and ultimately contributes to a more robust and informed scientific discourse. Remember, the pursuit of knowledge, even concerning the most enigmatic creatures, should always prioritize scientific integrity and responsible research practices. The strength of the research ultimately rests not in the definitive "proof" of a cryptid's existence but in the rigorous and transparent process undertaken to investigate its possibility.

The ethical dimensions of reporting cryptozoological research are particularly complex, demanding a nuanced approach that balances the pursuit of knowledge with the responsibility to avoid misrepresentation, sensationalism, and harm to the environment and its inhabitants. Transparency is paramount. This means making data collection methods, analytical processes, and all raw data readily accessible to the scientific community and, as appropriate, to the public. Data should be stored securely and managed ethically, respecting any potential privacy concerns related to individuals who may have contributed information. Openly sharing data allows other researchers to scrutinize findings, replicate methodologies, and contribute to the ongoing investigation. This collaborative aspect is crucial for building trust and enhancing credibility within the scientific community, combating the skepticism often associated with cryptozoological studies. The absence of transparent data handling can fuel suspicion and undermine the integrity of the research, regardless of the findings.

Accuracy is a non-negotiable element of ethical reporting. This necessitates the careful verification of all information before dissemination. Eyewitness accounts, often central to cryptozoological investigations, must be treated with particular caution. Researchers should employ multiple independent accounts whenever possible, employing rigorous cross-referencing and corroboration techniques. Similarly, photographic and video evidence should be rigorously analyzed for authenticity, considering potential distortions, manipulations, or misinterpretations caused by factors like lighting, angle, and resolution. The limitations of each data source should be explicitly acknowledged in the reporting, emphasizing the methodology used to account for potential biases or uncertainties. For example, if relying heavily on visual evidence, the resolution of the image and the conditions under which it was captured should be noted, along with any efforts taken to verify its authenticity, such as photo analysis by experts. Omitting this crucial detail can severely damage the credibility of the report.

The temptation to sensationalize findings, especially within a field like cryptozoology that often captures public imagination, must be strongly resisted. While generating public interest is vital for funding and awareness, doing so at the expense of scientific integrity is unacceptable. Exaggerating claims, manipulating data to create a more dramatic narrative, or deliberately omitting contradictory evidence are all serious breaches of ethical conduct. Hyperbole and speculative language should be strictly avoided; reporting should remain factual and evidence-based. For instance, instead of claiming "a monstrous creature was seen," a more responsible phrasing might be "a large, unidentified marine animal was sighted." Subtle differences in wording can significantly impact the perception of the report's objectivity and the credibility of the researcher. This commitment to responsible language applies equally to presentations, public lectures, and any media engagements. Ethical reporting should prioritize accurate communication over capturing headlines.

Another crucial ethical consideration lies in the potential impact of the investigation on the environment and its inhabitants. Cryptozoological investigations frequently involve fieldwork in sensitive ecosystems. Researchers must adhere to strict guidelines regarding environmental protection, obtaining any necessary permits and ensuring minimal disturbance to the habitat and its wildlife. Avoidance of habitat destruction, responsible waste management, and minimal interaction with the potential cryptid itself are all vital ethical aspects to consider. The study's potential effects should be comprehensively assessed before commencement and rigorously monitored throughout the investigation. For example, if the investigation focuses on an area of suspected cryptid activity, a baseline survey of the local flora and fauna before and after research is essential to assess any impact from researchers' presence. The ethical implications may extend beyond just environmental damage; the disruption of established ecosystem balances is also something to be carefully considered and minimized wherever possible.

Furthermore, the responsible handling of any potential discovery is ethically critical. If researchers believe they have discovered a new species, or have gathered evidence significantly supporting the existence of a previously unconfirmed cryptid, ethical considerations are paramount regarding the proper management and handling of the information. Public announcements should be approached cautiously, prioritizing collaboration with relevant scientific bodies and government authorities before any public communication. Premature or unverified announcements can lead to misinformation, speculation, and potential harm to the environment and the species itself. For instance, the excitement of a potential discovery might lead to a surge in tourism to an already fragile habitat, impacting local wildlife and ecosystems. Such uncontrolled access might even endanger the species or compromise future research efforts. In such instances, a phased and responsible approach to public communication is crucial, minimizing risks while ensuring that the discovery receives the attention it deserves.

Finally, respecting indigenous knowledge and cultural perspectives is crucial. Many cultures have long-standing oral traditions and narratives surrounding sea monsters, which often hold deep historical and cultural significance. Researchers should engage sensitively with local communities, respecting their traditional knowledge and practices. This might include consulting with community elders, incorporating local perspectives into the research, and acknowledging the indigenous names or interpretations of the potential cryptid. Failing to do so can represent a significant ethical lapse, potentially causing offense and disrespecting the cultural heritage of those communities whose territories may be involved in the investigation. Ethical engagement must be a cornerstone of any investigation, ensuring that the research is carried out in a respectful and culturally sensitive manner.

In conclusion, the ethical considerations involved in reporting cryptozoological findings are significant and multifaceted. Transparent data management, rigorous verification of accuracy, responsible avoidance of sensationalism, minimal environmental impact, prudent handling of potential discoveries, and respectful engagement with indigenous knowledge are all essential aspects of maintaining the integrity of the research and ensuring its responsible conduct. The pursuit of understanding these enigmatic creatures must always be guided by ethical principles that balance scientific inquiry with a commitment to environmental protection, cultural sensitivity, and responsible communication. The ultimate goal is not only the advancement of knowledge but also the preservation of the environment and the respectful integration of scientific research within existing cultural contexts.

10

Case Study: The Loch Ness Monster

The Loch Ness Monster, or Nessie, arguably the most famous of all cryptids, offers a rich tapestry of historical accounts and eyewitness testimonies, a body of evidence that has captivated and frustrated researchers for nearly a century. Analyzing this material requires a rigorous and critical approach, acknowledging inherent biases and limitations within the historical record. The earliest accounts, often steeped in folklore and legend, provide a fascinating, albeit unreliable, glimpse into the longevity of the Nessie myth. These tales, passed down through generations of Scottish communities, depict a creature of immense size, inhabiting the depths of the loch, often associated with mystical and supernatural elements. While these early stories lack the scientific rigor of modern accounts, they provide valuable insight into the cultural context surrounding the Nessie phenomenon, suggesting a deep-seated belief in its existence within the local population.

The transition from folklore to more formalized reporting occurred gradually, influenced by evolving perspectives on scientific observation and the growing accessibility of media. Early twentieth-century accounts, while still lacking the sophisticated technologies available today, represent a shift toward more structured documentation. Newspaper reports, letters, and personal journals from this period detail sightings of unusual objects in the loch, ranging from vague descriptions of disturbances on the water's surface to more specific accounts of elongated shapes and strange movements. These early accounts, though often anecdotal and lacking photographic evidence, form a crucial part of the historical record, showcasing the gradual emergence of the Nessie legend into the public consciousness. It is important to remember that the media of the era played a significant role in shaping public perception. Sensationalized reporting, fueled by a desire to capture reader interest, often exaggerated details and amplified the mystique surrounding Nessie, blurring the line between fact and fiction.

The mid-twentieth century saw a surge in eyewitness accounts, coinciding with the advent of more reliable methods of recording observations. The infamous photograph taken in 1934, famously dubbed the "surgeon's photograph," while later revealed to be a hoax, exemplifies the allure of visual evidence in bolstering claims of Nessie's existence.

This event, and the subsequent media frenzy, highlighted the power of visual representation in shaping public opinion and underscored the need for rigorous verification of photographic and video evidence in cryptozoological investigations. Subsequent eyewitness accounts from this era benefited from improved photographic technology, leading to a variety of images and film footage, although the quality and interpretability of this material often remain subjects of intense debate. Analysis of these images often reveals ambiguities, challenging the notion of definitive proof and highlighting the difficulties in discerning a unique cryptid from natural phenomena or misidentified creatures.

The latter half of the 20th century and the beginning of the 21st century witnessed a refined approach to collecting and analyzing evidence. The use of sonar technology, while still prone to errors, allowed for a more systematic search of the loch's depths, resulting in the detection of unusual underwater movements and large, undefined objects. While these sonar readings provided intriguing possibilities, they also demonstrated the challenges of interpreting underwater data. Sonar images often remain ambiguous, prone to misinterpretations due to factors such as water conditions, bottom topography, and the limitations of the technology itself. In addition, the use of advanced imaging technologies and underwater drones offer enhanced capabilities in exploring the depths of Loch Ness, generating high-resolution images and videos. Despite these technological advancements, however, interpreting the captured visuals and eliminating alternative explanations remains challenging.

A critical evaluation of eyewitness testimonies reveals numerous inherent biases. Confirmation bias, the tendency to interpret evidence in a way that confirms pre-existing beliefs, is a significant concern. Many eyewitnesses approach the search for Nessie with a preconceived notion of what the creature should look like, influencing their interpretation of observed phenomena. For example, a fleeting glimpse of a large, dark object in the water might be readily interpreted as Nessie, while a more cautious observer might attribute the sighting to a submerged log or other natural feature. This highlights the crucial role of unbiased observation and the need to consider alternative explanations for any observed phenomena.

Furthermore, the psychological phenomenon of pareidolia, the tendency to perceive meaningful patterns or images in random or ambiguous stimuli, further complicates the analysis of eyewitness accounts. The human brain is highly adept at recognizing patterns, often leading to the misinterpretation of natural formations, wave patterns, or even floating debris as evidence of a large, unknown creature. The lack of precise, consistent descriptions among eyewitnesses also undermines the reliability of the testimony. Accounts vary widely in terms of the creature's size, shape, and behavior, making it difficult to draw definitive conclusions. This inconsistency may reflect the limitations of observation in challenging conditions, such as poor visibility and limited time for observation. It may also indicate that multiple different phenomena are being misattributed to a single entity.

The cultural context within which these accounts are embedded also plays a critical role. The Loch Ness Monster has become a significant part of Scottish folklore and cultural identity. This cultural significance can influence eyewitness testimony, potentially leading to exaggeration or embellishment of observations in order to conform to existing beliefs or expectations. The economic impact of Nessie tourism also creates a potential conflict of interest, whereby the perpetuation of the legend may be economically advantageous, potentially motivating individuals to selectively emphasize or fabricate evidence supporting the existence of the creature.

In conclusion, a comprehensive review of historical accounts and eyewitness testimonies related to the Loch Ness Monster requires a multi-faceted approach. While the sheer volume of reported sightings cannot be dismissed outright, neither can their inherent limitations and biases. The lack of definitive, scientifically verifiable evidence necessitates a cautious and critical evaluation of all available material. The stories, though often compelling, must be examined within the context of their historical and cultural settings, acknowledging the potential for misidentification, misinterpretation, and even deliberate fabrication. The persistent mystery of Nessie underscores the challenges of cryptozoological research and the necessity of employing rigorous scientific methods, combined with a healthy dose of skepticism, in the pursuit of understanding these enigmatic creatures. Future investigations should focus on employing advanced technologies, collaborating with local communities, and developing more standardized methods of collecting and analyzing evidence to arrive at a more complete understanding of the phenomena surrounding Loch Ness.

The photographic evidence surrounding the Loch Ness Monster is, to put it mildly, underwhelming. While countless images purport to show Nessie, a rigorous analysis reveals a consistent pattern: ambiguity and lack of clarity. The most famous, the "surgeon's photograph" from 1934, initially fueled the legend but was later exposed as a hoax, a cleverly crafted miniature model. This incident serves as a cautionary tale, highlighting the ease with which images can be manipulated or misinterpreted, even unintentionally. Many other photographs depict blurry shapes, indistinct movements in the water, or fleeting glimpses of something that could be almost anything – a log, a wave pattern, a boat, or even a collection of waterfowl. The inherent limitations of early photographic technology, particularly in low-light and often turbulent water conditions, severely hampered the acquisition of clear, high-resolution images.

Even more recent photographs, captured with advanced digital cameras, often suffer from similar problems. The vastness of Loch Ness, its often murky waters, and the elusive nature of the supposed creature combine to create conditions where photographic evidence is rarely conclusive. Distant objects are difficult to resolve, and the lack of scale in many images makes it impossible to determine the true size of the objects photographed. Furthermore, the subjective nature of photographic interpretation comes into play. What one observer interprets as a large, serpentine creature, another might dismiss as a naturally occurring phenomenon or an artifact of the photographic process. The angle of the shot, lighting conditions, and even the psychological biases of the photographer can significantly influence the interpretation of an image.

The limitations of photographic evidence are exacerbated by the pervasive influence of pareidolia, the human tendency to perceive patterns in random data. Our brains are wired to seek meaning and order, and in the ambiguous context of a blurry image of Loch Ness, this tendency can lead to misinterpretations. A random arrangement of shadows and reflections might be interpreted as the head and neck of a large creature, whereas a more objective analysis might reveal a completely different explanation. The subjective nature of perception underscores the critical need for multiple, independent analyses of any photographic evidence, coupled with a thorough consideration of alternative explanations.

The application of sonar technology to the search for Nessie offers a potentially more objective approach, probing the depths of the loch without relying on surface observations. However, sonar data also presents significant challenges in interpretation. Sonar images are not direct representations of underwater objects; rather, they reflect the return of sound waves, which can be influenced by numerous factors, including water temperature, salinity, sediment density, and the presence of underwater currents. These factors can distort the sonar signal, leading to the creation of artifacts that could be misinterpreted as large objects or unusual movements.

The use of side-scan sonar, designed to produce a detailed image of the loch's bottom, has revealed numerous intriguing anomalies, such as large, undefined shapes or unusual underwater structures. However, the inability to visually confirm these anomalies hinders their interpretation. The shapes and structures depicted in sonar images may be anything from geological formations to submerged trees or wreckage. Without corroborating evidence from other sources, such as underwater video or physical inspection, it is impossible to definitively conclude that these sonar anomalies represent the Loch Ness Monster.

Furthermore, the limitations of sonar technology in terms of resolution and range also play a crucial role. Sonar images often lack the detail necessary to distinguish between different types of objects, hindering the ability to confidently identify a specific creature or structure. The detection range of most sonar systems is limited, meaning that objects beyond a certain distance may not be detected, or may be presented in a significantly distorted form. The use of multi-beam sonar, which offers improved resolution and coverage, has provided more detailed images of the loch floor, but still fails to provide definitive proof of Nessie's existence.

The challenges in interpreting both photographic and sonar data are further compounded by the lack of consistent and reliable eyewitness accounts. While numerous sightings have been reported, there's a significant lack of agreement on the creature's appearance, size, and behavior. This variability makes it difficult to reconcile different pieces of evidence and create a cohesive picture. A large, dark shape in the water could be many things, and without corroborating evidence, any given observation remains inherently inconclusive.

The integration of multiple data sources is essential for a more robust analysis. Combining photographic evidence with sonar data, alongside eyewitness accounts and historical records, could potentially paint a more complete picture. However, even with this multi-faceted approach, the inherent limitations of each data source must be acknowledged. The reliance on a single type of evidence, such as a photograph or sonar reading, is inherently flawed and risks leading to misinterpretations and unsubstantiated conclusions.

Moving forward, a more systematic and rigorous approach to collecting and analyzing evidence is crucial. The use of advanced technologies, such as high-resolution underwater cameras and remotely operated vehicles (ROVs), can provide more detailed images and videos of the loch's depths. The deployment of multiple ROVs simultaneously could enable triangulation of data, improving the accuracy and reliability of observations. Furthermore, the development of more sophisticated sonar systems with increased resolution and range would significantly improve the capacity to identify and characterize underwater objects.

However, technology alone is not sufficient. A rigorous scientific methodology is essential to minimize bias and ensure that evidence is interpreted objectively. This necessitates a multidisciplinary approach involving marine biologists, geologists, physicists, and experts in image analysis. Careful consideration of alternative explanations for any observed anomalies, including natural phenomena and known marine life, is crucial to avoid drawing premature conclusions. The application of robust statistical methods to analyze data from multiple sources can help to identify patterns and relationships that might otherwise be missed.

In conclusion, while both photographic and sonar evidence related to the Loch Ness Monster have generated significant interest and speculation, a critical assessment of these sources reveals considerable limitations. The inherent ambiguities in photographic images and the interpretive challenges associated with sonar data underscore the need for caution in drawing definitive conclusions. The absence of clear, unequivocal evidence, combined with the variability and often subjective nature of eyewitness accounts, necessitates a more rigorous and multidisciplinary approach to future investigations. The ongoing mystery of Nessie highlights the need for careful scientific methodology, an openness to multiple interpretations, and a recognition of the inherent difficulties in researching elusive cryptids in challenging underwater environments. The quest for Nessie continues, but the path forward must prioritize rigorous methodology and a balanced approach, incorporating both open-mindedness and a healthy dose of skepticism. Only through such a concerted effort can the search hope to yield conclusive and credible results.

The persistent enigma of the Loch Ness Monster necessitates a thorough exploration of alternative explanations, moving beyond the limitations of existing photographic and sonar evidence. A crucial aspect of this investigation involves considering misidentifications of known species inhabiting the loch. The waters of Loch Ness, while seemingly homogenous at the surface, harbour a surprising diversity of life, including various species of fish, eels, and invertebrates. These creatures, under certain conditions of light, distance, and water clarity, could easily be misinterpreted as something far more extraordinary.

Large eels, for example, possess an elongated, serpentine body shape that could, under less-than-ideal viewing conditions, resemble the commonly described form of Nessie. Their size, particularly older specimens, could contribute to misidentification, especially if only portions of the creature are visible. The European eel (*Anguilla anguilla*) is known to inhabit Loch Ness, and some specimens can attain remarkable lengths, exceeding two meters. Furthermore, their sinuous movements through the water, especially when disturbed, could easily be mistaken for the actions of a large, unidentified creature. The potential for misidentification is compounded by the generally poor visibility in the loch, limiting the detail available to observers.

Beyond eels, other inhabitants of Loch Ness could contribute to mistaken sightings. Large schools of fish, particularly salmon during their spawning runs, can create disturbances at the surface that might appear as the movement of a much larger creature. The coordinated actions of a large group of fish, combined with reflections of light off their scales, could amplify the visual effect, leading to misinterpretations. Similarly, seals, while uncommon in the loch, have been observed, and their sleek bodies and ability to surface suddenly and dive again swiftly could contribute to misidentification. Even playful otters, their sleek bodies moving rapidly through the water, could be misinterpreted as something far larger and more unusual, especially from a distance.

The role of geological formations in shaping the perception of the Loch Ness Monster must not be overlooked. The loch's bottom is characterized by a complex topography, including underwater ridges, valleys, and submerged vegetation. These features, when observed from the surface, especially in the often-murky waters, could easily create optical illusions, particularly in conjunction with the interplay of light and shadow. A submerged ridge, for instance, might appear as the back of a creature, while variations in water depth and the distribution of sediment could influence the apparent shape and size of objects underwater. The interaction of these geological formations with the water's surface – particularly its movement – could generate distortions that resemble a large, moving form.

The impact of environmental factors on visual perception cannot be overstated. The murky waters of Loch Ness, characterized by peat stains and fluctuating levels of suspended sediment, significantly limit visibility. The presence of waves, ripple patterns, and underwater currents can further distort the appearance of objects, creating an ambiguous visual environment prone to misinterpretation. Moreover, the interplay of light and shadow, particularly during dawn and dusk, can amplify the ambiguous nature of underwater sightings, making it challenging to discern the true nature of objects below the surface. The weather conditions, including fog and low-lying cloud cover, contribute further to the overall uncertainty in observations.

Moreover, the psychological aspects of observation play a vital role. Pareidolia, the human tendency to perceive patterns in random stimuli, significantly influences interpretations of ambiguous images and sightings. The human brain is naturally geared toward finding order and meaning, leading to the imposition of patterns onto random data. This phenomenon is especially pronounced in the context of the Loch Ness Monster, where the expectation of seeing a creature can significantly influence interpretation. Observers may readily attribute ambiguous forms to Nessie, unconsciously filling in the gaps to create a complete image that conforms to their preconceived notions. The inherent biases of the observer, their expectations, and even the social context of the observation can impact how they interpret what they see.

Beyond these purely visual aspects, we need to account for the potential impact of infrasound. While infrasound (sound waves below the range of human hearing) has gained traction as a theoretical means for communication within the aquatic world, particularly among marine mammals, its potential role in the interpretation of Loch Ness sightings has been largely overlooked. The generation of infrasound by large geological movements, seismic activity, or even weather phenomena could produce subtle vibrations that might be felt by observers without being directly heard. These vibrations, if misinterpreted, could contribute to an overall feeling of something unseen moving beneath the surface. Further complicating this potential, our own internal sensory mechanisms might be reacting without our conscious perception to other unexplained infrasound events in the loch itself, subconsciously informing our interpretations of vague visual observations.

Even the historical context of the Nessie legend significantly impacts modern interpretations. The decades-long proliferation of anecdotal accounts and imagery, often presented in a manner that reinforces the existing narrative, has likely amplified the tendency towards accepting ambiguous evidence as supporting Nessie's existence. This expectation bias, built up over time, makes it harder to objectively evaluate new claims or information. A rigorous analysis must, therefore, account for the influence of societal expectations and the ingrained cultural narrative surrounding the mystery.

Therefore, a comprehensive assessment of alternative explanations for Loch Ness Monster sightings necessitates a multi-pronged approach. It requires not just the analysis of photographic and sonar data, but also a deep understanding of the loch's ecology, its geological features, and the psychological influences shaping eyewitness accounts. Only through such a rigorous and multidisciplinary approach can we hope to disentangle the genuine unknowns from the misinterpretations and misidentifications that have fueled the enduring legend of Nessie. The pursuit of truth requires a steadfast commitment to objectivity, a willingness to explore all potential explanations, and a cautious approach to interpreting ambiguous evidence. Only with such rigor can the mystery of Loch Ness be adequately addressed.

The persistent mystery surrounding the Loch Ness Monster demands a rigorous, modern approach to investigation, moving far beyond the anecdotal evidence and blurry photographs that have dominated the narrative for decades. Fortunately, advancements in scientific technology and investigative methodologies offer unprecedented opportunities to re-examine this enduring enigma. One of the most promising avenues of research is environmental DNA (eDNA) analysis. This revolutionary technique involves collecting water samples from the loch and analyzing the DNA contained within to identify the species present. While it cannot definitively prove or disprove the existence of an unknown creature, eDNA analysis can provide a comprehensive inventory of the loch's biodiversity, revealing the presence of any unexpected or overlooked species. This information is crucial in eliminating potential misidentifications and establishing a clearer baseline of what actually inhabits the loch. The comprehensive dataset generated through eDNA analysis, covering both known and potential unknowns, forms a critical foundation for further investigations. By comparing the eDNA results with historical accounts and previous sightings, scientists might identify any inconsistencies or patterns that could shed new light on the mystery.

The limitations of eDNA analysis must, however, be acknowledged. The technique is primarily designed to detect the presence of DNA fragments, not the full organism. Degraded DNA, the common result of age or environmental stress, can be challenging to sequence accurately. The sensitivity of eDNA can also be influenced by water conditions such as temperature, currents, and the presence of pollutants. While it allows for the detection of species, the relative abundance or distribution of a given species is not necessarily accurately represented in the findings. The analysis requires considerable expertise and sophisticated laboratory equipment, increasing the cost and time needed for successful implementation. Furthermore, the absence of a specific DNA sequence doesn't necessarily imply the absence of the organism, due to the challenges of achieving a completely comprehensive analysis. Therefore, eDNA analysis serves as a valuable tool within a broader, multifaceted investigative approach, rather than a standalone solution.

Complementing eDNA analysis is the use of advanced underwater surveys. These surveys incorporate technologies that surpass the limitations of previous sonar and visual searches. Autonomous underwater vehicles (AUVs), equipped with high-resolution cameras and sonar systems, can systematically explore the depths of Loch Ness with an unparalleled level of detail. These unmanned vehicles can navigate the challenging underwater environment, reaching areas previously inaccessible to human divers or remotely operated vehicles (ROVs). The high-resolution cameras enable the capturing of detailed imagery of the loch's floor and its inhabitants, creating a visual record of the environment that can be subsequently analyzed for any unusual findings. The sonar data, with its capacity to penetrate the murky waters and map underwater topography, can reveal features that might be causing optical illusions or contributing to misinterpretations of previous sightings. By combining the visual data with acoustic data from sonar, investigators can obtain a more holistic understanding of the loch's environment. This systematic exploration is critical to ruling out any geological formations, submerged objects, or other environmental factors that could be mistaken for a large creature. Moreover, a comprehensive mapping of the loch floor can identify previously unknown features that could harbor unknown species, enhancing the overall biological survey.

Advanced imaging technologies are essential in evaluating existing evidence and analyzing new data. High-resolution image analysis techniques allow for a much more detailed scrutiny of previous photographs and video footage. These techniques can enhance the image quality, reduce noise, and identify subtle details that may have been overlooked in earlier assessments. Such enhancement might reveal identifying characteristics of a known species that could resolve a case of misidentification. Advanced image processing, utilizing sophisticated algorithms, is critical in evaluating the authenticity of images and video recordings, filtering out possible artifacts and manipulations. Furthermore, comparisons with existing databases of known species can provide more conclusive identifications. It may also reveal details that better inform interpretations of shape, size, and movement of any observed organism. This combination of careful analysis of existing records and the use of modern image processing technology enhances the reliability of visual evidence and helps minimize biases in interpretation.

Another promising area of modern investigation is the deployment of advanced sonar systems, extending far beyond the capabilities of earlier models. Side-scan sonar, for example, can create detailed images of the loch floor, revealing its topography and identifying any submerged structures or objects. Multibeam sonar provides even more comprehensive data, generating three-dimensional models of the underwater environment. This detailed mapping of the environment helps identify potential locations where an unknown creature might be hiding or frequent, or identify geological formations previously misinterpreted as a creature. In combination with high-resolution cameras and underwater positioning systems, multibeam sonar technology allows for a precision mapping of the loch's environment which contributes to a more systematic and rigorous survey of the loch.

The combined acoustic and visual data gathered will allow scientists to develop a much more detailed and accurate representation of the loch's environment. The use of sophisticated software and analysis techniques aids in isolating relevant data and identifying anomalies that warrant further investigation. Through rigorous application of advanced sonar, investigators can differentiate between naturally occurring structures and potentially unknown entities. By meticulously analyzing the resulting data, discrepancies and anomalies may indicate areas worthy of further, more focused exploration.

While modern investigative techniques offer substantial advancements, ethical considerations remain paramount. Loch Ness is a sensitive ecosystem, and any investigative efforts must be conducted in a manner that minimizes environmental disruption. The use of AUVs and other technologies requires careful planning to avoid damaging the habitat or disturbing the local wildlife. Environmental impact assessments must precede any major underwater surveys, with stringent protocols in place to mitigate potential harm. The researchers must follow strict guidelines to prevent disturbing the natural habitats of the various species. All data collected during the investigation must be treated with confidentiality and only used for scientific purposes, respecting any legal restrictions on access or publication of specific findings.

Open and transparent data sharing amongst scientific community is critical for collective verification and to facilitate ongoing investigations. The scientific integrity of any investigation is crucial, with the methodology and results subjected to peer review before publication. Transparency in data collection and analysis establishes credibility and builds public trust. A responsible and ethical approach ensures the long-term preservation of the loch's unique ecosystem while advancing our understanding of its mysteries.

In conclusion, the modern investigation of the Loch Ness Monster is a far cry from the anecdotal evidence and speculative interpretations of the past. The application of eDNA analysis, advanced underwater surveys, advanced imaging technologies, and sophisticated sonar systems offers a robust scientific framework for a fresh examination of this enduring mystery. By integrating these techniques, scientists can create a far more detailed and accurate picture of the loch's ecology and its inhabitants. However, it is imperative that this scientific rigor is coupled with a profound commitment to ethical practices to safeguard the delicate ecosystem of Loch Ness while striving for a deeper understanding of its secrets. The truth, whatever it may be, can only emerge from a diligent, methodical, and responsible approach to investigation. The enduring appeal of the Loch Ness Monster lies not just in the possibility of a hidden creature, but also in the broader quest for knowledge and the constant drive to refine our investigative approaches to unlock the secrets that nature holds. The mystery may or may not be solved, but the scientific journey in itself offers significant insights into modern ecological understanding and investigative methodology.

The accumulated evidence, while not providing a definitive answer to the question of the Loch Ness Monster's existence, significantly refines our understanding of the Loch Ness ecosystem and the limitations of interpreting historical accounts. The application of modern scientific methods has moved the investigation beyond speculation and into the realm of rigorous, data-driven inquiry. The initial results from eDNA analysis, while not revealing any unknown species unequivocally attributable to the "monster," have provided a far more comprehensive inventory of the Loch's biodiversity than previously available. This baseline understanding is crucial in ruling out numerous species previously proposed as potential candidates for Nessie, thereby narrowing the field of possibilities. Future eDNA studies should focus on increasing sample size and frequency, exploring different seasons and depths within the loch, and refining analytical techniques to enhance the sensitivity and accuracy of the results. The use of more advanced sequencing technologies will be critical in identifying even small fragments of DNA, thereby potentially uncovering trace evidence of elusive creatures. Furthermore, incorporating metabarcoding techniques will facilitate the identification of a broader range of organisms, improving the completeness of the biodiversity assessment.

The deployment of AUVs equipped with high-resolution cameras and multibeam sonar has revealed previously uncharted aspects of the Loch's underwater topography. High-resolution images have been generated, revealing the complexity of the lake bed, including various geological formations, submerged debris, and previously unknown underwater features. These findings provide vital context for interpreting historical sightings, demonstrating how seemingly unusual shapes and movements may be attributable to natural features or underwater objects. The use of AUVs for extended periods, covering a wider area of the loch, is vital for future surveys. This sustained observation will enhance our understanding of the Loch's dynamic environment and increase the likelihood of encountering any unusual phenomena or organisms. Furthermore, improved sensor technologies, such as hyperspectral imaging, could be integrated into future AUV missions. Hyperspectral imaging provides detailed spectral information that can help differentiate between various materials and organisms, potentially revealing subtle characteristics that would be otherwise overlooked.

Advanced sonar analysis has debunked several previously proposed explanations of sonar anomalies. The high-resolution multibeam sonar data has demonstrated that many of the intriguing sonar readings attributed to a large creature are in fact geological features or previously unidentified underwater objects. The comprehensive mapping produced through sonar significantly reduces the number of areas within the loch that warrant further investigation. Future sonar surveys should integrate real-time data processing and analysis capabilities to allow for immediate interpretation of the data during the survey. This real-time feedback will increase efficiency and enable the rapid identification of any anomalies requiring focused investigation. Moreover, incorporating other acoustic technologies, such as passive acoustic monitoring, would enhance the sensitivity of detection. Passive acoustic monitoring captures ambient underwater sounds, allowing for the detection of biological sounds, such as the vocalizations of unknown species.

The re-evaluation of historical photographs and video footage using advanced image analysis techniques has largely dismissed previously compelling evidence. Enhanced image processing techniques have revealed that many of the previously enigmatic images are actually misidentifications of known species or artifacts of photographic limitations and environmental factors. However, some images remain inconclusive, suggesting the need for more sophisticated image analysis techniques and a more nuanced understanding of image artifacts, Future research should focus on developing algorithms specifically designed to filter out artifacts commonly found in underwater imagery and to improve the accuracy of shape and size estimation. Collaboration with experts in computer vision and digital image processing will be crucial in achieving this goal.

The ethical considerations related to Loch Ness research must continue to be prioritized. The delicate ecological balance of the loch must be maintained, and any research must be carried out using minimally intrusive techniques. The development and implementation of stringent environmental impact assessments will be crucial in ensuring that research activities do not negatively impact the ecosystem. This includes carefully planning research routes and methodologies, minimizing disturbance to wildlife, and adhering to regulations concerning water quality and waste management. Transparent dissemination of research findings is also vital to foster public trust and promote collaboration amongst researchers and other stakeholders.

Beyond the specifics of the Loch Ness Monster investigation, this case study highlights the broader implications of using rigorous scientific methods in cryptozoological research. The application of eDNA analysis, advanced imaging, and sophisticated sonar technologies, combined with ethical considerations, offers a robust framework for future investigations into other alleged cryptids. The same methodological approach can be adopted for investigations in various aquatic environments worldwide, providing a standardized, scientific approach to the study of these elusive organisms.

The ongoing investigation of the Loch Ness Monster continues to serve as a powerful example of the importance of integrating scientific rigor with a sense of open-mindedness and critical thinking. While a definitive answer to Nessie's existence may remain elusive, the persistent pursuit of knowledge through innovative techniques and interdisciplinary collaboration drives us towards a richer understanding of the aquatic world and of the very nature of scientific inquiry. The potential discovery of a new species, or further refutation of existing claims, is of secondary importance to the invaluable lessons learned in the pursuit of this enduring mystery. The exploration is the real reward, revealing the wonders of nature, the intricacies of scientific investigation, and our enduring fascination with the unknown. The future of this research lies in collaborative efforts, embracing technological advancements and employing the most rigorous, ethical, and scientific approaches possible. Only through this sustained pursuit of knowledge can we truly appreciate the fascinating and complex natural world that surrounds us. The search for Nessie has evolved into a significant contribution to our understanding of limnology, underwater technologies, and the delicate balance of the aquatic environment, demonstrating that even the most seemingly outlandish investigation can yield significant scientific advancements. The quest for answers, however, remains an open invitation to future generations of researchers.

11

Case Study: The Gloucester Sea Serpent

The Gloucester Sea Serpent, a creature etched into the annals of maritime lore, presents a compelling case study in cryptozoology. Unlike the often-vague descriptions associated with other sea monsters, the Gloucester Serpent boasts a surprisingly detailed history, stemming from numerous eyewitness accounts spanning centuries. However, separating fact from fiction requires a meticulous dissection of these historical narratives, accounting for the inherent biases and limitations present in each account.

Early descriptions, dating back to the 18th century, paint a picture of a serpentine creature of immense size, often exceeding 100 feet in length. These accounts, frequently relayed through ship logs and newspaper articles, often emphasize the creature's sinuous form, its undulating movement through the water, and the terrifying spectacle it presented to sailors. Many accounts described a head unlike any known marine animal, sometimes likened to a horse's head, other times described as more reptilian, featuring a large jaw and prominent teeth. The discrepancy in descriptions, however, is a critical aspect that must be carefully considered when evaluating the reliability of these testimonies. Variations in lighting conditions, distance from the sighting, and individual perspectives all play a significant role in shaping how the creature was perceived and subsequently described. The emotional impact of encountering such a creature, especially in the often-harrowing conditions at sea, could also skew perceptions and distort memory.

The 1817 account by Captain Joseph Baker stands out in terms of its detail. Captain Baker claimed to have encountered the serpent near Gloucester, Massachusetts, noting its colossal size and the unique patterns on its skin. His account, while vivid, needs to be cross-referenced with other contemporaneous records, considering possible contextual influences such as existing sea serpent folklore and the potential for embellishment or exaggeration inherent in storytelling, especially within the context of 19th-century seafaring narratives. A similar consideration needs to be applied to other prominent accounts like that of the encounter described in the "Salem Gazette" of 1817. Sensationalized reporting was not uncommon, and the competitive nature of the early American press may have motivated accounts to be more dramatic than was truly warranted by the facts.

It's crucial to acknowledge the role of cultural context in shaping these eyewitness testimonies. The existence of established sea serpent myths and legends within the maritime community undoubtedly influenced how sailors interpreted their observations. A sighting of an unusual marine animal, however scientifically explainable, might be readily assimilated into the existing framework of pre-existing folklore, leading to accounts that reflect cultural expectations rather than objective observation. The influence of artistic representations of sea serpents, often exaggerated for dramatic effect, could further distort memories and recollections of sightings. In essence, the historical accounts form a tapestry woven from genuine observation, cultural narratives, and the potential for human error.

Beyond the written accounts, we must also examine the physical evidence, or lack thereof. There are scattered reports of purported remains, such as unusual vertebrae or skin fragments, attributed to the Gloucester Sea Serpent. However, thorough scientific examination of these specimens has typically yielded inconclusive or contradictory results. Many specimens initially claimed to be part of a sea serpent have later been identified as belonging to known species of marine animals, highlighting the importance of rigorous scientific verification in these cases. The difficulty in preserving organic material from the marine environment also casts doubt on the authenticity and reliability of many alleged specimens. The degradation of organic matter makes positive identification challenging, further complicating the analysis.

To determine the credibility of the sightings, we must consider factors that could contribute to misidentification. For instance, the undulating movement of large kelp forests or underwater currents could easily be misinterpreted as a serpentine creature. Large schools of fish, viewed from a distance, might also appear as a single, elongated form. Unusual marine phenomena such as bioluminescent displays or the movements of giant squid could also contribute to misinterpretations of the observed phenomena. The limitations of human perception at sea, coupled with variable visibility conditions, make accurate identification challenging.

The introduction of more advanced technologies in the mid-20th century brought with it a new era in marine exploration. The use of sonar, for example, revolutionized our ability to study the oceans' depths, offering a more scientific approach to investigate claims of unusual marine life. However, the interpretation of sonar data remains complex. Echoes from various underwater objects, geological formations, or even schools of fish can easily be misconstrued as larger, more organized structures, making it challenging to distinguish between potential cryptids and natural phenomena.

Furthermore, advances in underwater photography and videography offer another avenue for investigation. High-resolution images can help analyze the details of suspected sightings. However, problems of water clarity and distance continue to limit image quality. Moreover, the sheer expanse of the ocean makes systematic visual surveys extremely challenging.

The Gloucester Sea Serpent, despite the wealth of historical accounts and eyewitness testimonies, remains a fascinating and unresolved enigma. To approach this case responsibly, a balance is needed between acknowledging the compelling nature of the historical records and maintaining scientific rigor in the evaluation of the available evidence. The challenge lies in separating factual observations from embellishment, cultural influence, and misinterpretation. Ultimately, while we may never definitively determine the existence of the Gloucester Sea Serpent, its enduring legend serves as a powerful reminder of the mysteries that still reside in our oceans and the importance of applying a robust scientific methodology to unravel them.

The study of the Gloucester Sea Serpent showcases the inherent challenges and complexities of cryptozoology, illustrating the need for comprehensive data collection, critical analysis, and a clear understanding of the limitations of historical accounts and eyewitness testimonies. Even in the absence of concrete evidence, the investigation continues to inspire further exploration of the marine environment, driving technological advancements and our understanding of the still-mysterious depths of the ocean. The ongoing quest to understand this potential cryptid highlights the value of maintaining a robust and interdisciplinary approach, always incorporating rigorous scientific methods while simultaneously keeping an open mind to the possibilities that exist within the unexplored realms of the ocean.

The accounts of the Gloucester Sea Serpent, while varied, offer a fascinating lens through which to examine the challenges of cryptozoological investigation. A consistent theme across multiple descriptions is the creature's immense size, often estimated to be over 100 feet in length. This colossal scale immediately sets it apart from any known marine animal. However, the reliability of these size estimations hinges on several factors, not least of which is the inherent difficulty of accurately judging size at sea, particularly when dealing with a rapidly moving and potentially elusive creature. The distance between the observer and the serpent, the visibility conditions (affected by factors like weather, light, and water clarity), and the emotional state of the observer all contribute to potential inaccuracies. A creature of such immense size would undoubtedly be a terrifying sight, potentially affecting the observer's ability to accurately assess its dimensions.

The descriptions of the serpent's physical form are similarly inconsistent. While the serpentine nature is a constant, the specifics of its head and body vary widely. Some accounts describe a horse-like head, others a more reptilian visage with a powerful jaw and prominent teeth. Such discrepancies highlight the limitations of eyewitness accounts and the influence of pre-existing beliefs and expectations. Sailors familiar with sea serpent folklore might be more inclined to interpret ambiguous shapes and movements in line with established cultural narratives, potentially distorting their perception of the creature's features. The absence of standardized methods for recording sightings in the 18th and 19th centuries further exacerbates these inconsistencies. Lack of precise measurement tools, combined with the chaotic nature of encounters at sea, makes it exceptionally difficult to achieve reliable descriptions.

Furthermore, the accounts frequently mention unusual coloration and skin patterns. Some descriptions refer to a mottled or striped appearance, while others mention specific color variations. Again, the reliability of these descriptions is questionable due to the subjective nature of color perception under varying light conditions. The effects of light refraction in water further complicate the accurate identification of color and pattern. The psychological impact of observing such a creature should also be considered. Fear and stress can profoundly affect an individual's perception and memory, potentially leading to exaggerated or distorted accounts of the creature's physical attributes. It's crucial to distinguish between genuine descriptions and the potential impact of fear-induced distortions in the recollection of details.

The difficulty in objectively assessing the reliability of physical descriptions is compounded by the lack of concrete physical evidence. While there have been reports of alleged remains – vertebrae, skin fragments, etc. – scientific examination has, in almost all cases, yielded inconclusive or contradictory results. The harsh conditions of the marine environment significantly hamper the preservation of organic material. Decomposition rates are high, and the process of decomposition itself can alter the appearance of remains, making identification extremely challenging. Furthermore, the possibility of misidentification of existing marine life fragments as belonging to a cryptid is ever-present, necessitating rigorous scrutiny and cross-referencing with existing biological data.

Even the most seemingly straightforward aspects of the physical descriptions prove difficult to interpret. For instance, the consistent mention of the serpent's "undulating" movement. While this aligns with the expected movement of a large serpentine creature, it's essential to consider alternative explanations. Large kelp forests, for example, can exhibit a similar undulating movement in strong currents. The swaying motion of such underwater vegetation, when viewed from a distance or in poor visibility conditions, could easily be misinterpreted as the movement of a large aquatic animal. Similarly, the aggregation of large schools of fish, especially when viewed from above the waterline, might present a seemingly elongated and serpentine shape, potentially leading to misidentification.

The lack of high-quality photographic or video evidence further hinders the analysis of the creature's physical description. The technology to capture detailed underwater imagery simply didn't exist during the period of the most prolific sightings. While early photographic technology did exist, its limitations, coupled with the challenging environment of the ocean, would have made obtaining high-resolution images almost impossible. This absence of concrete visual evidence leaves the analysis reliant primarily on subjective eyewitness accounts.

The comparative analysis with known species is also crucial. While no known marine animal perfectly matches the descriptions of the Gloucester Sea Serpent in terms of size and morphology, it's essential to explore the possibility of misidentification of existing marine life. The sighting of an unusual specimen of a known species, observed from a distance or in poor visibility, could easily be interpreted as a cryptid. Large eels, certain types of whales (at a distance), or even unusually large specimens of less-known species could potentially be misinterpreted. A robust comparative analysis, using detailed biological data and morphological comparisons, is necessary to rule out misidentifications.

It's also crucial to understand the limitations of our perception, particularly in the challenging environment of the open ocean. Distance, visibility, and the emotional impact of the encounter all play a significant role in shaping how eyewitnesses perceive and remember a sighting. Human perception is inherently fallible; even under optimal conditions, accurate estimations of size, color, and morphology can be challenging. In the turbulent and unpredictable environment of the sea, these limitations are significantly magnified.

In conclusion, while the historical accounts of the Gloucester Sea Serpent paint a vivid picture of a colossal and terrifying creature, the analysis of physical descriptions reveals significant challenges in separating factual observations from misinterpretations, exaggerations, and the influence of pre-existing cultural beliefs. The absence of concrete physical evidence and the limitations of human perception in the challenging marine environment further complicate the investigation. Despite these challenges, the investigation of the Gloucester Sea Serpent serves as a valuable case study in cryptozoology, highlighting the importance of rigorous scientific methodology, careful consideration of contextual factors, and a balanced approach that acknowledges both the compelling nature of the historical narratives and the inherent limitations of the available evidence. The pursuit of answers in this enigmatic case continues to drive technological advancements and our understanding of the vast and still largely unexplored mysteries of the deep ocean. The quest to unravel the truth behind the Gloucester Sea Serpent is far from over, and its continuing legacy encourages ongoing research and improved investigative techniques within the field of cryptozoology.

Given the inconsistencies and limitations inherent in the historical accounts of the Gloucester Sea Serpent, a crucial step in any objective analysis involves exploring possible alternative explanations. This involves meticulously examining the possibility of misidentification of known marine species, coupled with a thorough understanding of how environmental conditions can distort human perception and contribute to embellished or inaccurate accounts.

One of the most critical aspects of this investigation involves a comparative analysis with known marine life. While no currently known species perfectly matches the reported size and description of the Gloucester Sea Serpent, several species could potentially account for some, if not all, of the sightings. Large basking sharks, for instance, can reach impressive lengths, and their relatively docile nature and the way they often feed near the surface could lead to misinterpretations, especially if viewed from a distance or in conditions of poor visibility. Their dark coloration and the occasional presence of parasitic growths could be perceived as unusual markings, contributing to the enigmatic descriptions of the serpent's appearance. Similarly, giant squid, although typically inhabiting deeper waters, occasionally surface, and their immense size and unusual morphology could easily be misinterpreted by observers unfamiliar with these creatures. The fleeting nature of these sightings, coupled with the often chaotic conditions at sea, only adds to the potential for misidentification.

Furthermore, the behavior of certain whale species needs to be considered. Several species of whales, particularly those with elongated bodies, might, under specific circumstances, be mistaken for a serpentine creature. A group of whales swimming in a coordinated manner, particularly at a distance, could create the illusion of a single, massive, undulating form. The spray produced by a whale's blowhole, viewed from a distance, might even be misinterpreted as a creature's head or unusual markings. The subjective nature of eyewitness accounts, combined with the lack of sophisticated observation tools during the era of the most prolific sightings, makes the possibility of whale misidentification a compelling area of investigation.

Another aspect to consider is the role of less-commonly observed, yet large, marine creatures. The ocean's depths remain largely unexplored, and many species, particularly those inhabiting the twilight zone or deeper waters, remain poorly understood. The possibility that some of the Gloucester Sea Serpent sightings refer to an unusual specimen of a known but infrequently encountered species is not easily dismissed. Such an occurrence, coupled with the limitations of early observation methods, could readily contribute to the exaggerated descriptions and inconsistent accounts found in historical records. A comprehensive review of all known marine species, both common and rare, is required to adequately explore this possibility.

Beyond misidentification of specific species, environmental factors play a critical role in shaping eyewitness accounts. Ocean currents, for instance, can create unusual wave patterns and movements that, when viewed from afar, might be interpreted as a large creature moving beneath the surface. The refraction of light in water can further distort the perception of size, shape, and color, leading to inaccurate descriptions. Conditions such as fog, rain, or low light levels would only exacerbate these perceptual distortions. The interplay of these environmental variables with the inherent limitations of human perception highlights the challenges involved in objectively evaluating historical accounts.

The impact of psychological factors must also be taken into account. The sighting of an unusually large or unfamiliar marine creature at sea would undoubtedly evoke a strong emotional response, potentially leading to heightened anxiety, fear, and even panic. Such states profoundly affect memory and perceptual accuracy, potentially leading to embellishments and distortions in the recounting of events. The stress of the situation, coupled with the inherent uncertainty of an encounter with the unknown, can significantly impact the reliability of eyewitness accounts. It is essential to consider the psychological impact of the sighting to accurately assess the credibility of the reported details.

Moreover, the influence of pre-existing cultural narratives cannot be overlooked. Sea serpent legends are deeply ingrained in maritime folklore, and the expectation of encountering such a creature could unconsciously influence the interpretation of ambiguous sightings. Sailors familiar with these tales might be more prone to interpreting naturally occurring phenomena or unusual marine life as evidence of a sea serpent, effectively projecting pre-existing beliefs onto their observations. Understanding the cultural context surrounding these sightings is essential to separating factual observations from the influence of pre-existing expectations and biases.

Finally, the absence of concrete physical evidence is a significant constraint in the investigation. Despite numerous reports of alleged remains, the lack of verifiable specimens further complicates efforts to identify the Gloucester Sea Serpent. The harsh marine environment, which rapidly degrades organic material, is a major challenge in preserving specimens for scientific analysis. Furthermore, the possibility of misidentification of existing marine life debris as belonging to a cryptid presents a significant hurdle in verifying reported findings.

In conclusion, while the accounts of the Gloucester Sea Serpent present a compelling narrative, a robust investigation necessitates a careful evaluation of potential alternative explanations. The misidentification of known marine species, the influence of environmental factors on perception, the impact of psychological responses, and the role of pre-existing cultural beliefs all need to be thoroughly considered. The lack of physical evidence and the limitations of early observational techniques further complicate any definitive identification. By critically evaluating these alternative explanations, we can move toward a more nuanced understanding of the Gloucester Sea Serpent enigma, acknowledging both the intriguing historical accounts and the challenges inherent in investigating cryptic marine phenomena. The pursuit of understanding continues to emphasize the need for rigorous methodology, combining open-mindedness with critical analysis in the ongoing quest to unlock the secrets of the deep ocean. The ongoing investigation serves as a testament to the enduring mysteries of the marine world and highlights the importance of continued research and methodological advancements in the field of cryptozoology.

The exploration into alternative explanations strengthens the scientific approach to cryptozoological investigations, emphasizing the importance of empirical evidence and critical thinking in differentiating between credible claims and instances of misidentification. The lack of conclusive proof, rather than debunking the possibility of a cryptid, highlights the need for further exploration, improved technology, and a more refined understanding of the ocean's still largely uncharted depths. The Gloucester Sea Serpent, therefore, transcends its status as a single case study and acts as a significant example in the broader discourse of cryptozoology, pushing the boundaries of scientific investigation and our understanding of the unexplored wonders within the oceans of our planet.

The historical accounts of the Gloucester Sea Serpent, while captivating, suffer from a significant lack of contemporary, scientifically rigorous data. Applying modern investigative approaches reveals both promising avenues and frustrating limitations. Historical research, for instance, forms a crucial foundation. Newspaper archives, ship logs, and personal journals from the relevant periods offer invaluable glimpses into the sightings, allowing researchers to piece together a timeline and geographic distribution of reported encounters. However, the inherent biases and limitations of these primary sources must be carefully considered. Many accounts are anecdotal, relying on secondhand information or subjective interpretations of events. The lack of standardized reporting methods during that era—the absence of standardized photographic techniques, precise location data, or detailed descriptions following a consistent format—hampers accurate comparison and analysis across different accounts. Furthermore, the cultural context of the time, heavily influenced by existing sea serpent lore and a penchant for sensationalism in the press, adds further complexity to disentangling fact from fiction.

To counter the shortcomings of historical accounts, modern investigative tools offer potential advantages. Sonar technology, for example, could theoretically detect large, underwater objects in the areas where the Gloucester

Sea Serpent was reportedly sighted. However, the sheer size of the relevant marine areas, the limitations of older sonar systems, and the potential for misinterpretations of geological formations or schools of fish significantly constrain the effectiveness of this approach. Moreover, the unpredictable nature of the ocean environment and the fleeting nature of the alleged sightings further reduce the chances of a successful detection. Similarly, the use of remotely operated vehicles (ROVs) and autonomous underwater vehicles (AUVs) presents possibilities. These advanced technologies enable visual inspection of the seabed and the water column, potentially identifying unknown large marine life. Yet, the cost, logistical challenges, and time involved in underwater exploration significantly limit the feasibility of conducting exhaustive searches across such vast regions.

The use of genetic analysis represents another promising but challenging avenue. While the lack of physical specimens prevents direct analysis of the Gloucester Sea Serpent itself, researchers could investigate the possibility of unknown species through the analysis of environmental DNA (eDNA). By collecting water samples from the reported sighting locations and sequencing the DNA contained within them, it might be possible to identify traces of unknown large organisms. The technology is still developing, however, and the complexities of interpreting eDNA results from such diverse marine ecosystems remain considerable. Environmental factors like currents and degradation processes can significantly affect the presence and viability of eDNA, further hindering the analysis.

Photographic and video evidence, while rare for historical sightings, becomes increasingly vital in modern investigations. High-resolution underwater cameras, both remotely operated and deployed from vessels, can capture detailed imagery, assisting in the identification of marine life. However, even advanced technologies are limited by the conditions of the underwater environment, often hampered by turbidity, lack of light penetration, and the unpredictable movements of marine creatures. The chance of capturing clear images or videos of a large, elusive creature remains extremely low.

Furthermore, analyzing historical accounts through a modern lens of behavioral ecology can offer valuable insights. For instance, considering the migratory patterns of known large marine animals, the areas frequented by specific species, and their typical behavior in relation to human activities can help eliminate certain possibilities. This approach involves mapping historical sighting locations and comparing them with the known distribution ranges of various marine species. It allows for a systematic evaluation of possible misidentifications and helps identify potential areas for future focused investigation.

The limitations, however, are equally significant. The lack of precise location data in historical accounts complicates the task of accurately comparing sighting locations with the ranges of known marine animals. Furthermore, our understanding of marine species behavior remains incomplete, especially for deep-sea creatures, meaning that we might not fully understand the factors driving unusual behavior or surfacing events.

Moreover, a crucial element in any investigative approach is establishing a rigorous methodology. This involves defining specific research questions, outlining clear investigation protocols, and employing standardized data collection and analysis techniques. This enhances objectivity and reduces bias, leading to more credible findings. Establishing a clear framework for evaluating evidence, including criteria for assessing the reliability of historical accounts, photographic evidence, and any other relevant data, further minimizes subjective interpretation.

The ethical implications of the investigation should also be considered. Unnecessary disturbances to the marine environment during underwater surveys must be avoided. The protection of marine ecosystems and the ethical treatment of any marine life encountered should be paramount. Researchers must follow stringent guidelines and obtain necessary permits before conducting any fieldwork.

In conclusion, while modern investigative approaches offer promising tools for investigating the Gloucester Sea Serpent enigma, the lack of contemporary, high-quality data presents significant challenges. The limitations of historical accounts, coupled with the logistical and technological hurdles involved in underwater exploration, highlight the difficulty of conclusively addressing the question of the creature's existence. Nevertheless, a rigorous and ethical application of modern scientific methods, combined with critical analysis of available data, can contribute to a more nuanced understanding of the historical reports, even if a definitive answer remains elusive. The Gloucester Sea Serpent case, therefore, serves as a valuable case study in the application of scientific principles to cryptozoological investigations, emphasizing the importance of careful methodology, meticulous data analysis, and a constant awareness of the limitations of both historical and contemporary scientific tools.

The pursuit of knowledge in this realm demands an iterative process of hypothesis testing, data collection, and critical evaluation, continuously refining our understanding of the ocean's mysteries and our capacity to investigate them. The search for the truth, even when shrouded in uncertainty, pushes the boundaries of scientific inquiry and underscores the enduring allure of the unexplored depths. The enduring fascination with the Gloucester Sea Serpent, despite the lack of definitive proof, highlights the enduring human curiosity and the relentless pursuit of understanding the enigmatic wonders of the natural world. The very limitations of the investigation serve to underscore the vastness of our oceans and the much we still have to discover within their depths. Ultimately, the story of the Gloucester Sea Serpent may never be fully resolved, but the ongoing pursuit of answers contributes significantly to our collective knowledge of marine biology and the evolution of investigative techniques within cryptozoology.

The Gloucester Sea Serpent case, while frustrating in its lack of conclusive evidence, offers a valuable lesson in the challenges and rewards of investigating historical cryptozoological claims. The inherent limitations of historical accounts – the absence of standardized reporting methods, the prevalence of anecdotal evidence, and the influence of cultural biases – severely restrict the possibilities for definitive proof. Eyewitness testimonies, while compelling in their narrative power, are often subjective and susceptible to misinterpretation or exaggeration. The lack of precise location data, detailed physical descriptions, and corroborating evidence from multiple, independent sources makes it incredibly difficult to reconcile different accounts and arrive at a cohesive picture.

Furthermore, the technological limitations of the past significantly hinder our ability to analyze historical sightings with modern scientific tools. While sonar technology, ROVs, and AUVs offer unprecedented opportunities for underwater exploration, their application to historical events is inherently retrospective. We cannot retroactively deploy these technologies to the specific locations and times of the reported sightings. Therefore, any attempt to verify or refute these accounts using modern technology must rely on extrapolations and educated guesses, inherently increasing the margin of error.

Even the relatively recent advancements in genetic analysis, specifically environmental DNA (eDNA) analysis, present significant challenges. While eDNA holds immense potential for detecting unknown species through water samples, the degradation of DNA over time, the complexities of marine ecosystems, and the sheer volume of water to be sampled make it highly unlikely that traces of the Gloucester Sea Serpent, if it ever existed, would be detectable today. The environmental factors, such as currents and biological processes that degrade DNA, represent a significant hurdle to overcoming in any such retrospective analysis.

The search for photographic or video evidence, equally elusive, highlights the technological limitations of the past. The absence of high-quality imagery from the periods of peak sighting activity significantly hampers any attempts at visual identification. While modern underwater cameras offer exceptional capabilities, the task of revisiting historical locations and capturing any lingering evidence is monumental, given the vastness of the ocean and the often imprecise locations provided in historical accounts.

The application of modern behavioral ecology to historical data, while seemingly promising, also encounters considerable difficulties. Mapping historical sighting locations and comparing them to the known distributions of large marine animals may reveal patterns or inconsistencies. However, the inherent imprecision of historical location data, coupled with incomplete knowledge of deep-sea animal behavior, makes definitive conclusions challenging. Our current understanding of marine life, especially the behavior of deep-sea species, is incomplete, creating significant limitations when trying to interpret historical accounts through a contemporary ecological lens.

Beyond the methodological challenges, ethical considerations must be paramount in any modern investigation. Any attempts at underwater exploration must be conducted responsibly, minimizing environmental impact and adhering to strict regulations. Researchers should prioritize the protection of marine ecosystems and avoid unnecessary disturbance to marine life. The pursuit of knowledge should never come at the expense of environmental integrity. The meticulous planning of any fieldwork, therefore, is paramount in any such endeavor.

The Gloucester Sea Serpent case, therefore, serves as a stark reminder of the limitations of investigating historical cryptozoological claims. While modern investigative techniques offer advanced tools and perspectives, the lack of contemporary data, the challenges of retrospective analysis, and the inherent uncertainties of interpreting historical accounts necessitate a cautious and nuanced approach.

Despite the difficulties, the investigation remains valuable for several reasons. Firstly, it highlights the importance of rigorous scientific methodology, even when dealing with seemingly elusive phenomena. The establishment of clear research questions, standardized data collection protocols, and objective analysis techniques are crucial, regardless of the subject matter. Secondly, the case demonstrates the ongoing evolution of investigative tools and techniques within cryptozoology. The advancement of technologies like sonar, ROVs, AUVs, and eDNA analysis continuously expands our ability to explore and understand the marine environment, even if their application to historical cases is challenging.

Thirdly, the Gloucester Sea Serpent case underscores the interplay between history, culture, and science in shaping our understanding of the natural world. The historical accounts reflect not only potential sightings of an unknown creature but also the cultural context of their time, including prevailing beliefs, biases, and the media's influence on public perception. A comprehensive investigation requires careful consideration of both the scientific and socio-cultural factors that contributed to the narrative. Examining historical accounts, therefore, provides valuable insights into the historical perspectives and sociocultural factors that have shaped the perceptions of the oceans and their inhabitants.

Fourthly, the enduring fascination with the Gloucester Sea Serpent, despite the lack of definitive proof, speaks to the enduring human curiosity about the unknown and the mysteries that lie beneath the ocean's surface. The story itself, regardless of its ultimate resolution, motivates a constant push towards developing more sophisticated investigative tools and a more comprehensive understanding of the marine environment. This pursuit ultimately benefits not only cryptozoology but also the broader field of marine biology, fostering innovation and driving scientific advancement.

In conclusion, while the Gloucester Sea Serpent enigma might remain unsolved, the investigation continues to offer valuable insights. The case study serves as a powerful illustration of the complexities involved in investigating historical cryptozoological claims. It underscores the importance of rigorous scientific methodology, careful consideration of ethical implications, and a realistic appraisal of the limitations inherent in both historical and contemporary investigative tools. The ongoing research, while not guaranteeing definitive answers, contributes significantly to our understanding of marine life, the evolution of investigative techniques, and the enduring power of human curiosity. The ongoing research and exploration in this field, therefore, promise to yield significant contributions to our understanding of the ocean's mysteries and its many hidden denizens, even if the exact nature of the Gloucester Sea Serpent remains elusive. The legacy of this case study lies not solely in solving the mystery, but in advancing the methodologies and technologies that will help to uncover the truth behind future cryptozoological inquiries. The pursuit continues, fueled by the constant desire to unravel the ocean's secrets and expand our collective knowledge of the planet's extraordinary biodiversity.

12

Case Study: The Giant Squid

The colossal squid, *Architeuthis dux*, a creature of myth and legend for centuries, only began to yield its secrets relatively recently. Before the advent of modern deep-sea exploration techniques, our understanding was shaped by fragmented accounts, often shrouded in the fog of folklore and exaggerated seafaring tales. These early encounters, while lacking the rigor of modern scientific observation, provide a fascinating glimpse into the evolution of our perception of this enigmatic cephalopod. The historical accounts offer a rich, albeit unreliable, tapestry woven from the threads of eyewitness testimony, artistic renderings, and secondhand narratives passed down through generations.

Early accounts frequently conflated the giant squid with other large marine creatures. Descriptions often blur the lines between giant squid, colossal squid, and even large octopus species, reflecting the limited knowledge and observational tools available at the time. The sheer size and the infrequent sightings often led to hyperbole, transforming already impressive creatures into monstrous leviathans of the deep. Sailors, faced with the inexplicable and often terrifying reality of encountering a massive, unknown creature in the inky blackness of the ocean's depths, understandably relied on existing cultural frameworks and pre-existing narratives to interpret their experiences. These frameworks, steeped in folklore and myth, often amplified the already extraordinary nature of the encounter, contributing to the enduring mystique surrounding the giant squid.

Consider the accounts from the 16th and 17th centuries, chronicled in ship logs and personal diaries. Many lacked detailed scientific descriptions, instead favoring vivid, often sensationalized, accounts designed to capture the awe and terror inspired by the creatures. The lack of standardized reporting and the influence of cultural biases heavily shaped the narrative, frequently casting the giant squid as a fearsome sea monster, a threat to sailors and ships. These early chronicles, while lacking scientific accuracy, nevertheless laid the groundwork for future investigations, hinting at the existence of a creature far exceeding the size and scale of any known cephalopod. These initial glimpses, however inaccurate, ignited the curiosity and fueled the investigation that would ultimately lead to a better understanding.

The artistic representations of the period also reflect this early understanding. Paintings and woodcuts depict giant squid in highly stylized and often exaggerated forms, further reinforcing the mythical and monstrous image. These images, often intended to capture the imagination rather than depict reality, contributed to the enduring perception of the giant squid as a creature of legend. The artistic license taken by artists and illustrators further compounded the challenge of separating fact from fiction in the existing historical record. While providing visual glimpses of early encounters, these images often fail to accurately represent the true biology and morphology of the creature.

The nineteenth century brought a shift, with the emergence of more structured scientific inquiry. While the lack of advanced technologies like submersibles and underwater cameras persisted, naturalists and scientists began to approach sightings with a more systematic approach, though still hindered by technological constraints. Improved communication allowed for a wider dissemination of information, leading to the collection of more comprehensive accounts, although the potential for biases and misinterpretations remained substantial. The lack of standardized methodology for documenting observations, however, continued to hinder progress.

The famous accounts of giant squid encounters during the 19th century are often described in the context of their respective cultural landscapes. The prevalence of certain narratives and the exaggeration of size can be understood as a product of the social and cultural beliefs of the time. In many cases, sailors' stories, influenced by existing myths and the pressure to embellish tales of the sea, significantly contributed to the ambiguity that surrounded early accounts. Distinguishing actual sightings from creative embellishments often requires a profound understanding of the historical, social and cultural contexts in which these reports emerged.

The scientific community also struggled to reconcile accounts of the giant squid with existing biological knowledge. The sheer size of the creatures challenged the prevailing understanding of cephalopod biology. The gradual accumulation of specimens, mostly dead or decaying parts washed ashore, provided some empirical evidence, but the fragmented nature of the material hindered a full understanding of the animal's anatomy and behavior. The process of developing a scientifically valid understanding was a gradual evolution, built on years of piecing together fragmented evidence, with substantial delays resulting from the difficulties in observing and collecting samples.

The 20th century witnessed a gradual improvement in the understanding of giant squid, mainly due to advancements in deep-sea exploration technology. The development of improved underwater cameras, remotely operated vehicles (ROVs), and submersibles allowed for direct observation of giant squid in their natural habitat, though such encounters remain exceedingly rare. These observations offered a significant step toward verifying and refining earlier accounts. The advancement of technology allowed for direct imaging and the collection of less fragmented biological samples, leading to a more accurate understanding of the anatomy and behavior of these elusive creatures.

Even with these advancements, challenges persist. The deep-sea environment remains largely unexplored, and the creatures' natural habitat, characterized by immense pressure and darkness, presents significant logistical challenges. Acquiring complete specimens remains difficult, and much about their behavior, reproductive strategies, and population dynamics remains unknown. The rarity of sightings continues to limit the accumulation of data, underscoring the magnitude of the challenge in fully understanding this incredible species.

The historical accounts and early encounters with giant squid, although often lacking precision and sometimes laced with exaggeration, are invaluable as they represent our initial efforts to grapple with the existence of a creature so extraordinary that it seemed to defy the boundaries of reality. These early observations, though flawed by the constraints of the time, form the foundation upon which our current, more scientifically rigorous, understanding has been built. The path from mythical sea monster to a scientifically recognized species demonstrates the importance of combining historical investigation with technological advancement, cultural understanding, and rigorous scientific methodology in the field of cryptozoology. The continued exploration of the deep ocean, and the ongoing application of new technologies, promises to further enrich our understanding of the giant squid and its place in the intricate tapestry of marine life. The journey from folklore to fact, therefore, remains ongoing, demonstrating the constant evolution of knowledge in the pursuit of truth, however elusive.

The colossal squid, *Architeuthis dux*, presents a fascinating paradox: a creature of immense size and striking appearance, yet one shrouded in mystery due to its deep-sea habitat and elusive nature. While our understanding has significantly improved in recent decades, thanks to technological advancements, many aspects of its biology and behavior remain enigmatic. Unraveling these mysteries requires a multi-faceted approach, combining historical data with modern scientific methods.

Focusing on its physical characteristics, the colossal squid stands out as one of the largest invertebrates on Earth, rivaled only by its close relative, the giant squid. However, while both species are gigantic, crucial differences exist. The colossal squid typically possesses a shorter mantle length than the giant squid, but compensates for this with significantly longer arms and tentacles, making it arguably heavier overall. Reports suggest colossal squid can reach mantle lengths of up to 2-4 meters, with total lengths, including arms and tentacles, potentially exceeding 10 meters

– some extreme claims even reaching far beyond this. Verification of these extreme measurements remains challenging due to the difficulty in capturing and accurately measuring complete specimens. The existing data largely comes from incomplete specimens or from fleeting observations in its deep-sea habitat.

One of the most striking features of the colossal squid is its exceptionally large eyes. These eyes, reportedly the largest in the animal kingdom, are thought to be an adaptation to the deep-sea environment, maximizing light gathering capabilities in the near-total darkness of its habitat. The sheer size of these eyes suggests a crucial role in prey detection and predator avoidance in the dimly lit depths where it resides. Studies on the structure of the eyes suggest remarkable adaptations for low-light vision, capable of detecting the faint bioluminescence emitted by potential prey or the shadows of approaching predators. The detailed structure and function of these eyes are still areas of ongoing research, potentially revealing valuable insights into the evolutionary adaptations of deep-sea creatures.

The colossal squid's arms and tentacles are equipped with formidable hooks, a key distinction from the giant squid, which possesses primarily suckers. These powerful hooks, arranged in rows along the arms and tentacles, serve a vital purpose in capturing and securing prey. Analysis of the morphology and biomechanics of these hooks suggests they are remarkably strong and effective for gripping and holding onto struggling prey. The size and arrangement of these hooks vary along the arms and tentacles, suggesting specialized roles in different aspects of prey capture and manipulation. Microscopic analysis has revealed a unique composition of the hook material, enhancing their durability and grip in a harsh, high-pressure environment.

Another distinctive feature of the colossal squid is its relatively small beak compared to its body size, in contrast to the giant squid's proportionately larger beak. The function and strength of the beak in relation to prey consumption remain an area of active research. While both beaks are undoubtedly powerful tools for breaking down prey, the smaller beak of the colossal squid might suggest differences in prey selection or feeding strategies compared to the giant squid. It is possible that the colossal squid utilizes its hooks more extensively for prey capture, relying less on a powerful beak to initially subdue its food.

The coloration of the colossal squid is reported as varying shades of reddish-brown, though the exact coloration can be highly variable, potentially depending on factors such as age, sex, and depth. The coloration likely serves a camouflage purpose in the deep-sea environment, helping to conceal it from both predators and prey. The exact pigment responsible for this reddish-brown hue, and its potential role in countershading or other forms of camouflage, requires further research.

The behavior of the colossal squid remains largely uncharted territory, due to the difficulties of observing these elusive creatures in their natural environment. However, based on limited observations, evidence suggests a solitary lifestyle. Direct observation of live specimens is extremely rare, with most of our understanding coming from specimens found dead or dying. Therefore, extrapolating behavioral patterns from these limited observations necessitates considerable caution.

Limited evidence suggests that colossal squid may be ambush predators, employing their large eyes and hooks to capture prey in the deep sea. Potential prey items could include various deep-sea fish, smaller squid, and even other cephalopods. The analysis of stomach contents from recovered specimens offers some insights into their diet, however, the limited sample sizes restrict the strength of such conclusions. Further study is needed to fully understand their dietary habits and ecological role within the deep-sea ecosystem.

The reproductive biology of the colossal squid remains largely unknown. The lack of observations of mating behavior and the scarcity of gravid females limits our understanding of their reproductive strategies. Detailed studies on the anatomy of reproductive organs are currently providing some hints, yet much remains to be uncovered regarding spawning behaviors, egg development, and parental care.

The deep-sea environment poses significant challenges to studying colossal squid, making their research exceptionally complex and resource-intensive. The extreme pressures, darkness, and vastness of their habitat limit the scope of direct observation, necessitating the use of advanced technologies such as remotely operated vehicles (ROVs) and autonomous underwater vehicles (AUVs) for underwater research. These technologies, however, are costly and require significant logistical planning. Furthermore, the inherent rarity of these animals makes encounters, even with the assistance of these advanced tools, exceptionally rare.

The ongoing research efforts to understand the colossal squid encompass diverse approaches. These include analyzing recovered specimens, deploying deep-sea cameras to directly observe their behavior, using advanced sonar technology to detect their presence in the water column, and incorporating genetic analyses to better understand their evolutionary history and relationships with other cephalopods. All these various approaches complement each other, yielding a more comprehensive and nuanced understanding.

The colossal squid's existence underscores the vast unknown of the deep ocean. Its study not only provides crucial insights into the biodiversity and adaptations of deep-sea life but also highlights the ongoing need for technological advancements and innovative research methodologies in the exploration of our planet's least-understood ecosystems. The journey to fully understanding this enigmatic creature promises to uncover further marvels and complexities within the abyssal depths. The pursuit of knowledge about the colossal squid is an ongoing process, reminding us of the vastness of the natural world and the constant evolution of our understanding of it. The future holds the promise of unveiling more of its secrets, through continued dedication to deep-sea exploration and meticulous scientific investigation.

The ongoing quest to understand the giant squid, *Architeuthis dux*, has been revolutionized by advancements in deep-sea exploration technology and a surge in research initiatives. For decades, our knowledge was largely based on fragmented specimens washed ashore, often damaged and incomplete. These offered tantalizing glimpses into the creature's morphology, but provided little insight into its behavior or ecological role. The advent of remotely operated vehicles (ROVs) and autonomous underwater vehicles (AUVs), however, has dramatically altered the landscape of giant squid research. These remotely controlled submersibles, equipped with high-definition cameras and specialized lighting systems, can now venture into the deep ocean, accessing the giant squid's natural habitat. This capability allows for direct observation of living specimens in their environment, providing unprecedented opportunities to study their behavior, hunting strategies, and interactions with other organisms.

Early ROV deployments focused primarily on visual observation and video recording. The resulting footage, often grainy and fleeting due to the challenging conditions, provided nonetheless crucial information. Observations revealed unexpected aspects of giant squid behavior, challenging some long-held assumptions. For instance, contrary to some early depictions, many observations suggest that giant squid are not necessarily aggressive hunters relentlessly pursuing prey, but may employ ambush tactics, relying on camouflage and their impressive size to surprise victims. Such observations highlight the importance of moving beyond anecdotal evidence and incorporating direct, real-time observations to reconstruct the giant squid's natural history.

Further advancements in underwater imaging technology have refined our understanding of giant squid. Higher-resolution cameras, combined with specialized lighting techniques to penetrate the abyssal darkness, enable far more detailed observation of the squid's anatomy, including subtle details of their skin texture, chromatophore patterns, and the delicate structures of their tentacles and suckers. This detailed imaging is crucial for understanding the role of camouflage and communication in the giant squid's life. The development of sophisticated 3D imaging techniques, capable of capturing three-dimensional models of the squid in their natural environment, adds another layer of detail, allowing scientists to analyze their body proportions and movement patterns with unprecedented accuracy.

Beyond visual observation, ROVs and AUVs have also enabled the collection of biological samples from living giant squid. This is a significant breakthrough, as it allows for genetic analysis, providing deeper insights into the giant squid's evolutionary relationships, population structure, and genetic diversity. Analyzing the squid's DNA can resolve long-standing questions about the population structure and distribution of giant squid around the globe, possibly elucidating variations in size, morphology and behavior between different populations. Genetic research also offers an essential tool to debunk folklore and differentiate true *Architeuthis dux* from other deep-sea cephalopods that might have been mistaken for it.

Sonar technology has emerged as another critical tool in the search for giant squid. By using acoustic signals to detect the presence of large marine organisms in the water column, sonar surveys can help pinpoint potential giant squid habitats and predict locations for ROV deployments. The development of advanced sonar systems with improved resolution and sensitivity allows for a much more precise detection of large moving objects in the deep ocean, significantly increasing the chances of encountering a giant squid. By mapping the distribution of giant squid in relation to environmental factors such as water temperature, salinity, and depth, scientists can gain a much better understanding of the environmental preferences and habitat requirements of these mysterious creatures.

The integration of multiple research approaches, combining ROV observations, genetic analyses, sonar surveys, and detailed anatomical studies of recovered specimens, is critical for building a comprehensive picture of the giant squid's life. This holistic approach allows scientists to cross-validate findings obtained from different methodologies, reducing uncertainties and building stronger conclusions about the giant squid's biology and behavior. For instance, sonar data indicating the presence of large organisms in a specific area can be used to inform the deployment of ROVs, which can then provide detailed visual observations and sample collection, further corroborated by genetic studies of collected tissues.

The scientific community's understanding of giant squid remains incomplete, but the pace of discovery has accelerated dramatically in recent years. Ongoing research efforts continue to unravel the enigmas surrounding their distribution, behavior, reproduction, and interactions with the deep-sea ecosystem. Future advancements in underwater technology, coupled with the continued collaborative efforts of scientists from various disciplines, promise to unveil further wonders of this colossal cephalopod and expand our knowledge of the deep ocean's biodiversity. The quest to understand the giant squid serves not only as a fascinating case study in cryptozoology, but also provides crucial insights into the mysteries of the deep sea, and the challenges inherent in studying life in extreme environments.

The study of giant squid also highlights the importance of interdisciplinary collaboration in scientific research. Marine biologists, geneticists, oceanographers, and engineers are working together, leveraging each other's expertise and technologies to address the many challenges associated with studying this elusive animal. This collaborative approach is crucial not only for maximizing efficiency in research efforts but also for fostering creative problem-solving, leading to breakthroughs in understanding and technologies. The ongoing exchange of ideas and information among different research groups accelerates the progress of knowledge, promoting a faster, more efficient, and thorough understanding of the natural world.

Furthermore, ethical considerations play an increasingly important role in giant squid research. Minimizing disturbance to the deep-sea ecosystem and avoiding harm to the animals during observations are paramount. Strict guidelines for ROV operations and sample collection are in place to ensure minimal impact on the animals and their habitat. The development of less intrusive observation techniques, such as bioluminescent sensors that can remotely track giant squid without direct contact, are also under investigation. This emphasizes the growing awareness of the need for responsible and sustainable research practices in the pursuit of scientific knowledge.

The ongoing pursuit of knowledge about giant squid is a testament to humanity's enduring curiosity and our determination to explore the unknown corners of our planet. As technology continues to advance and research efforts intensify, we can expect to witness a further evolution in our understanding of this magnificent creature and its role in the vast and enigmatic deep-sea ecosystem. The continuing investigations underscore the importance of respecting the natural world and conducting scientific research with care, ethics, and a commitment to long-term sustainability. The information gained not only unveils more of nature's wonders but also enhances the appreciation of the challenges and rewards inherent in the scientific process. The giant squid, in many respects, continues to serve as a symbol of the unexplored depths of our oceans and the vast expanse of knowledge yet to be uncovered. The story of giant squid research is far from over – it is an ongoing testament to scientific curiosity, technological innovation, and the persistent human drive to comprehend the mysteries of the natural world.

The enduring fascination with the giant squid has, unfortunately, fostered a fertile ground for misconceptions and myths. These inaccuracies, often rooted in historical accounts and embellished by popular culture, have significantly hampered a clear understanding of *Architeuthis dux* and its true nature. Separating fact from fiction requires careful examination of these ingrained beliefs and a robust comparison with the increasingly accurate scientific data emerging from recent research.

One of the most persistent myths surrounding the giant squid is its monstrous size and aggressive nature. Early accounts, often recounted by sailors with limited scientific training, dramatically inflated the size of encountered squid, describing gargantuan creatures capable of sinking ships with their tentacles. These narratives, fueled by a natural fear of the unknown and the often perilous conditions of seafaring, transformed the giant squid into a mythical kraken, a fearsome beast capable of immense destruction. While the giant squid is indeed a large creature, capable of reaching impressive lengths, the reality is far less dramatic.

Scientific measurements from recovered specimens and observations from ROVs consistently show that while exceptionally large, they are significantly smaller than the colossal proportions depicted in seafaring legends. Moreover, recent observations suggest a more passive hunting strategy, utilizing ambush tactics rather than the relentless pursuit frequently depicted in folklore. The tendency to exaggerate encounters, amplified over time through storytelling and embellishment, contributed to the enduring image of the giant squid as an insatiable, aggressive predator.

Another prevalent misconception revolves around the giant squid's habitat and distribution. The deep-sea environment, inaccessible for centuries, fueled speculation about its widespread presence and seemingly impossible habitats. Many accounts placed giant squid in waters far beyond their known range, suggesting they inhabited almost any depth or region of the ocean. Modern research, however, paints a more nuanced picture. While the giant squid's distribution appears to be quite widespread, with confirmed sightings in various locations across the globe, it's not ubiquitous. Its distribution is far more constrained and linked to specific environmental factors such as ocean currents, water temperature, and the presence of suitable prey.

This understanding stems from dedicated sonar surveys, genetic analysis correlating specimen locations, and systematic mapping of deep-sea ecosystems, revealing the preferred habitats of giant squid, refuting the notion of their presence in virtually all oceanic depths and regions. The fact that many deep-sea habitats remain largely unexplored contributes to the lingering belief that giant squid are much more common than scientific data suggests. The vastness of the unexplored deep ocean continues to fuel speculation and allows for the persistence of unfounded claims about the animal's widespread nature.

The lack of sufficient live observations in the past also fueled a wealth of misinterpretations regarding the giant squid's behaviour. Many narratives depicted the squid as attacking ships, ensnaring them in its tentacles, and dragging them to the depths. These dramatic tales were often embellished for entertainment or to emphasize the perils of seafaring. However, modern deep-sea exploration employing ROVs has provided crucial insight into their more subtle behaviour. Observations reveal a creature that is not relentlessly aggressive, but instead employs ambush tactics, relying on its camouflage and size to surprise prey. This behaviour is far more nuanced and less dramatic than the depictions found in many historical accounts and fictionalised narratives. The initial lack of direct observation naturally led to significant misinterpretations, as the stories were passed down through generations, further exaggerating the actual behavior of the animals.

The visual appearance of the giant squid has also been the subject of numerous exaggerations and misconceptions. Early descriptions often depicted creatures with enormous eyes, massive tentacles adorned with razor-sharp hooks, and a menacing beak capable of tearing through ships. While the giant squid does indeed possess significant anatomical features, these aspects were often significantly overstated in folklore. The size and shape of its beak and eyes, while impressive, have been subjected to considerable exaggeration. The tentacles, while incredibly powerful, do not possess the sharp, hook-like structures often depicted in fantasy. These differences highlight the difficulty of accurate observation in the challenging deep-sea environment and the natural tendency to embellish accounts that already possess an element of the extraordinary. The lack of precise and high-quality imagery further contributed to the propagation of such inaccuracies.

The colouration of giant squid is another area where common perceptions diverge sharply from scientific understanding. Many early accounts described the giant squid in a variety of lurid colours, often attributed to the deep-sea environment's perceived influence on the animal's pigmentation. While some colour change is possible through chromatophores, which allow the squid to change its skin tone for camouflage or communication, the diversity and vibrancy of colors described in many fictional accounts far exceed the limited observed spectrum. The colour change ability is often far more subtle than portrayed in many narratives. Modern, high-resolution images from ROVs show a colour palette far more subdued than the highly imaginative descriptions found in traditional accounts.

Moreover, the supposed intelligence of the giant squid has been frequently misunderstood. The lack of comprehensive data in the past fuelled speculation of near-human or even superhuman intelligence. While cephalopods are indeed known for their intelligence, attributing supernatural cognitive capabilities to the giant squid is a gross overstatement. While undoubtedly intelligent creatures, capable of complex behaviours, the giant squid's cognitive abilities are likely within the normal range expected of cephalopods, not on a scale exceeding even the most intelligent of mammals.

The persistence of these myths about giant squid highlights the challenges inherent in studying elusive creatures inhabiting harsh, difficult-to-access environments. The deep sea, a vast and largely unexplored realm, provided a fertile ground for speculation, allowing myths and misconceptions to persist for centuries. However, the advancement of deep-sea exploration technology, combined with rigorous scientific methodologies, is steadily dispelling these inaccuracies and providing a more accurate and nuanced understanding of the giant squid. The continuing research endeavors not only provide valuable insights into the life of this extraordinary creature but also serve as a powerful illustration of the scientific process and the ongoing quest to uncover the secrets of the deep ocean.

The unraveling of these myths represents a critical step towards appreciating the unique and fascinating realities of the giant squid and its place within the deep-sea ecosystem. The careful consideration of scientific findings, alongside the critical examination of historical accounts and folklore, is crucial in obtaining a fully comprehensive understanding of this majestic creature. The work remains ongoing, with much still to be discovered, but the journey away from misconception and toward a scientifically sound knowledge base is a journey well worth undertaking.

The giant squid's enigmatic nature, coupled with its elusive deep-sea habitat, has firmly cemented its place within the annals of cryptozoology. Its very existence, once relegated to the realm of sailors' tales and maritime legends, now stands confirmed by scientific evidence, yet the creature continues to occupy a unique space within the field. The discrepancy between the mythical kraken and the scientifically documented *Architeuthis dux* highlights the complexities inherent in bridging the gap between folklore and fact. This discrepancy, however, isn't simply a matter of dismissing old wives' tales; it reveals the vital role cultural narratives play in shaping our understanding of the natural world, and how those narratives can both hinder and ultimately aid in scientific discovery.

The historical accounts of giant squid encounters frequently depict battles of epic proportions, with colossal creatures capable of wreaking havoc upon unsuspecting vessels. These narratives, often embellished over time through repeated retellings, contribute significantly to the creature's mystique. The sheer size and power attributed to these mythical squids, far exceeding the dimensions observed in actual specimens, speak to the human tendency to amplify the extraordinary, transforming a real animal into something monstrous. Yet, even in these exaggerated accounts, a kernel of truth persists, a testament to the animal's awe-inspiring size and the undeniable power it wields within its deep-sea domain.

The deep-sea environment itself adds another layer to the giant squid's cryptozoological significance. The abyssal plains, with their crushing pressures, near-total darkness, and extreme temperatures, were largely uncharted territory for centuries. This inaccessibility fueled speculation and conjecture, making the deep ocean a breeding ground for legends and myths, many of which involved gigantic, terrifying creatures lurking in the inky blackness. The giant squid, dwelling within this mysterious realm, naturally became an iconic symbol of the unknown, its very existence a validation of the terrifying power attributed to the ocean's depths. Even with modern technology capable of exploring the deep sea, areas remain inaccessible, sustaining the lingering sense of mystery and potential for undiscovered wonders within this extreme environment.

Furthermore, the limitations of early observation technologies significantly impacted the perception of the giant squid. Before the advent of remotely operated vehicles (ROVs) and deep-sea submersibles, observations were largely limited to glimpses from ships, often resulting in distorted or incomplete images. These fleeting glimpses, coupled with the challenges of observing a creature in its natural habitat, readily led to inaccurate estimations of size, behaviour, and even physical characteristics. The distorted perspectives and subjective interpretations inherent in these encounters inevitably led to the amplification of unusual traits, further fueling the imaginative narratives that continue to shape our understanding of the creature.

However, the very existence of these myths and legends offers a valuable resource for researchers. Analyzing historical accounts, even those heavily embellished, can provide clues to the animal's geographic distribution, behavioural patterns, and even physical characteristics. By comparing historical accounts with contemporary scientific findings, researchers can glean a deeper understanding of the animal's true nature. The biases and cultural influences embedded within these narratives need to be carefully considered, but when carefully analyzed, they provide a fascinating window into past perceptions of the natural world and highlight the evolution of scientific understanding.

The giant squid's unusual anatomical features also play a role in its cryptozoological appeal. Its enormous size, relative to other squid species, its unique camouflage abilities, and its powerful tentacles contribute to its mystique. Early accounts, lacking the detail provided by modern imaging techniques, often attributed fantastical features to the creature, intensifying its aura of otherworldliness. While scientific understanding has clarified some of these exaggerated features, the actual anatomy of the giant squid remains impressive enough to inspire awe and wonder.

Moreover, the limited number of observed specimens has fueled speculation. The infrequent encounters with live giant squid, coupled with the difficulties of retrieving intact specimens from the deep sea, have contributed to the perception of the creature as exceedingly rare and elusive. This scarcity has enhanced its legendary status, turning it into a symbol of the mysterious and the unknown. The difficulty of observation is, however, entirely understandable given the extreme habitat the animal occupies and the technological challenges associated with such deep-sea exploration. As more observations are made using sophisticated underwater equipment, a more accurate picture is emerging, but the very challenges of observing these animals ensure that the element of mystery will linger.

Interestingly, the giant squid's position within cryptozoology isn't solely about debunking myths. The creature acts as a potent reminder of the limitations of our knowledge, the vastness of the unexplored ocean, and the potential for surprises within the natural world. It forces us to critically examine our assumptions, to question what we believe we know, and to remain open to new discoveries. The ongoing research into giant squid serves not only to refine our understanding of the species itself but also to improve our methodology for investigating other elusive deep-sea creatures.

In conclusion, the giant squid's role in cryptozoology is multifaceted and complex. It represents a compelling case study of how folklore, limited observations, and the challenging deep-sea environment can intertwine to create a creature shrouded in myth and legend. However, the creature is not merely a product of myth; it is a remarkable animal whose existence, although now confirmed by science, still holds significant mystery. The continued investigation of giant squid isn't just about unraveling the myths surrounding it; it's about uncovering the truths of the deep sea and refining our methods for exploring the uncharted territories of our planet. The ongoing research into this remarkable creature continues to push the boundaries of scientific understanding and to remind us of the vast unknown that lies beneath the ocean's surface. The persistent fascination with the giant squid, born from a mixture of fear, wonder, and the mystery of the deep, is a testament to the enduring power of the unknown and the enduring human desire to explore the limits of our world. The giant squid remains a powerful symbol, not only of the deep sea but also of the ongoing exploration of the unknown, and the essential interplay between myth, legend, and scientific discovery.

13

The Future of Sea Cryptid Research

Technological advancements are rapidly reshaping the landscape of sea cryptid research, offering unprecedented opportunities to investigate these elusive creatures. The limitations of past research, often hampered by rudimentary equipment and subjective eyewitness accounts, are being overcome by sophisticated technologies that provide more objective and detailed data. This shift promises a more rigorous and scientifically robust approach to the study of sea cryptids, bridging the gap between folklore and scientific understanding.

One of the most significant advancements is in underwater imaging. Early explorations relied on limited visibility and crude photography, resulting in blurry images and often inaccurate interpretations. The development of high-resolution underwater cameras, equipped with powerful lighting systems capable of penetrating the deep ocean's darkness, allows for detailed observation of marine life in its natural habitat. Remotely Operated Vehicles (ROVs) and Autonomous Underwater Vehicles (AUVs) further enhance this capability, providing maneuverability and the ability to explore previously inaccessible areas of the ocean. These technological marvels allow for close-up observation of marine organisms, capturing high-definition video footage and still images for detailed analysis. The improvement in image resolution is crucial; blurry images, often the result of poor visibility or distance, led to misinterpretations and fueled speculative accounts in the past. High-resolution images provide a level of clarity that allows for accurate species identification and the detection of subtle features that may indicate the presence of an unknown organism.

Furthermore, the integration of advanced sonar systems significantly improves the ability to detect and locate marine life. Traditional sonar systems often lacked the resolution to distinguish between different types of marine organisms, but modern sonar technologies, including multibeam sonar and side-scan sonar, offer high-resolution mapping of the seafloor and the water column. These systems can detect objects and organisms at considerable distances, even in areas with low visibility, providing a broader picture of the underwater environment. This is particularly useful in tracking the movements of large or elusive creatures, potentially identifying areas where sea cryptids might be more likely to occur. The development of advanced acoustic sensors, capable of picking up sounds and bioluminescent signals, enhances this ability further, opening up another dimension to underwater detection and surveillance.

Beyond imaging, the field of environmental DNA (eDNA) analysis is proving to be a revolutionary tool in sea cryptid research. eDNA refers to the genetic material shed by organisms into their environment, such as skin cells, scales, and feces. By collecting water samples and extracting the eDNA, researchers can identify the species present in a particular area, even if the organisms themselves are not directly observed. This technique is especially valuable in the investigation of elusive creatures that are rarely sighted. The advantage of eDNA lies in its ability to detect the presence of organisms that may be scarce or elusive, enhancing the ability to identify unknown organisms through DNA sequencing and comparative analysis with existing databases. While a match might not immediately identify a cryptid, it might reveal a previously unknown species whose characteristics may shed light on legendary sea creatures.

The potential of eDNA is further enhanced by advances in genomic sequencing technology. The rapid speed and decreasing cost of genome sequencing allow for quick and efficient identification of species from even small amounts of eDNA. This speed and efficiency are crucial in the pursuit of cryptic organisms, allowing researchers to rapidly analyze large volumes of environmental samples in order to detect the presence of unknown organisms in a reasonable time-frame. The identification of previously unknown genetic sequences might point towards the existence of new species, some of which might align with previously described sea cryptids.

The integration of artificial intelligence (AI) is transforming the analysis of both underwater images and eDNA data. AI-powered image recognition systems can process vast amounts of visual data far more quickly and efficiently than human analysts, identifying potential cryptids that might be missed by human eyes. These systems are continuously learning and improving their accuracy through machine learning, making them increasingly effective tools for identifying both known and unknown species. By feeding these systems with images of known species, they can develop a robust and accurate capacity to differentiate between known marine life and potentially new species. This capacity to process large volumes of data is particularly crucial in detecting rare or infrequent sightings of potential cryptids.

Similar AI applications are being developed for the analysis of eDNA data. AI algorithms can analyze complex genomic datasets, identifying patterns and relationships that may not be immediately obvious to human researchers. These algorithms can assist in comparing unknown sequences to extensive databases, potentially revealing the identity of cryptic organisms or suggesting the presence of entirely new lineages. AI is also streamlining the process of filtering through extensive datasets and generating potential leads that are then investigated more thoroughly by human researchers. The combination of AI-driven analysis with human expertise offers a powerful approach to unlocking the mysteries of elusive species, bridging the gap between vast datasets and meaningful interpretation.

Furthermore, advancements in sensor technology are enhancing our ability to gather data from the deep ocean. Bioacoustic sensors are becoming increasingly sophisticated, capable of detecting a wide range of sounds produced by marine organisms, including those produced by unknown creatures. These sounds can provide valuable clues about the size, behaviour, and location of elusive species. Advances in autonomous underwater vehicles (AUVs) allow for extended exploration of the deepest and most inaccessible parts of the ocean, collecting data from areas that were previously unreachable. The combination of these technologies allows for a comprehensive approach to data collection that combines observations from multiple modalities, increasing the likelihood of identifying cryptic species.

The ethical considerations surrounding sea cryptid research are equally important. Researchers must ensure that their investigations do not harm the marine environment or the creatures they are studying. This necessitates a careful and responsible approach to data collection, minimizing the impact on marine ecosystems. This is especially critical in the deep ocean, a fragile and sensitive environment that can be easily disturbed. The use of non-invasive techniques, such as eDNA sampling and remote sensing, is crucial to this ethical approach, ensuring the wellbeing of the environment and the animals within it.

Ultimately, the future of sea cryptid research depends on embracing and integrating these technological advancements. By combining rigorous scientific methodology with cutting-edge technology, researchers can move beyond speculation and anecdotal accounts towards a more objective and comprehensive understanding of the ocean's mysteries. The integration of these technologies ensures a more robust approach, moving away from subjective interpretations and offering greater accuracy in species identification and behavioral observations. The path forward necessitates a balance between scientific rigour and open-mindedness, embracing the potential for new discoveries while maintaining ethical practices. The pursuit of these elusive creatures is not just about finding them; it is about expanding our knowledge of the marine world and enhancing our understanding of the biodiversity that lies beneath the waves. The future holds exciting possibilities for the field, potentially revealing species that were previously thought to exist only in legend.

The pursuit of sea cryptids, those elusive creatures residing in the ocean's depths, demands a collaborative effort transcending traditional scientific boundaries. The complexity of the marine environment, coupled with the inherent challenges of underwater research, necessitates a multidisciplinary approach, bringing together expertise from various fields to ensure comprehensive and robust investigations. This collaborative spirit is paramount, fostering a synergy between different perspectives and methodologies, leading to a more complete understanding of both the legendary creatures and the marine ecosystems they inhabit.

Central to this collaborative approach is the partnership between cryptozoologists and marine biologists. Cryptozoologists, often possessing extensive knowledge of historical accounts, folklore, and anecdotal evidence related to sea monsters, provide a crucial historical and cultural context. They bring to the table a deep understanding of the legends, often including details about creature descriptions, behaviours, and locations, which can guide more targeted investigations. Their expertise in analyzing eyewitness accounts, interpreting ambiguous imagery, and assessing the credibility of various sources contributes significantly to the investigative process.

Marine biologists, on the other hand, offer essential scientific grounding, providing crucial expertise on marine ecology, taxonomy, and the general biology of marine organisms. Their knowledge of known species is invaluable in differentiating potential cryptids from misidentified or poorly understood creatures. This collaboration ensures that potential sightings are evaluated not only in the context of historical accounts, but also within the framework of current scientific understanding of marine biodiversity.

The partnership expands beyond these two core disciplines. Oceanographers contribute their invaluable expertise in understanding ocean currents, water temperature gradients, and seabed topography, which can inform the search for potential cryptid habitats. Their understanding of the physical characteristics of the marine environment is vital in assessing the plausibility of reported sightings and in designing effective search strategies. For example, knowledge of specific currents could help to predict potential migration routes of a large, unknown marine creature, guiding the allocation of research resources. Similarly, information on specific seabed features, like deep-sea trenches or hydrothermal vents, can point towards areas where unique and undiscovered species might thrive. The integration of oceanographic data into cryptid investigations adds a layer of scientific rigor, enhancing the precision and objectivity of the research.

Geologists and geophysicists contribute further to this interdisciplinary approach by providing insights into the geological history and composition of the ocean floor. Their knowledge of tectonic plates, seamount formations, and historical geological events can shed light on potential habitats and the evolution of unique marine ecosystems. The existence of certain geological formations, for instance, might create isolated pockets of unique environments that could harbor unique or previously unknown species. By understanding the geological context, researchers can better target their investigations to regions that are more likely to support the existence of unique and elusive marine fauna. This integration of geological knowledge helps to move beyond simple biological considerations and to understand the wider environmental context in which a potential cryptid may exist.

The expertise of engineers and technologists is also critical in overcoming the technical challenges inherent in deep-sea research. They play a crucial role in designing and building the sophisticated equipment needed for underwater exploration, including remotely operated vehicles (ROVs), autonomous underwater vehicles (AUVs), and advanced sonar systems. These technological advancements, as detailed in earlier chapters, are vital for collecting high-quality data and images from the ocean depths, allowing for a more detailed and scientifically sound investigation of potential cryptid sightings. Without the contributions of engineers and technologists, much of the modern research methodology discussed earlier would be impossible to implement effectively.

Furthermore, the collaboration extends to experts in genetics and bioinformatics. The analysis of environmental DNA (eDNA) now plays a crucial role in identifying potential cryptids. Genetic sequencing allows for the identification of unknown organisms through their DNA, even without a direct visual observation. The expertise of geneticists is needed to interpret genetic data, compare it against existing databases, and make inferences about the possible relationship between unknown DNA sequences and potential cryptids. Bioinformaticians, on the other hand, are vital in analyzing the vast amounts of genetic data generated, employing computational tools to identify patterns and relationships that might otherwise be missed.

Moreover, the involvement of experts in data analysis and statistics is critical in the objective assessment of evidence. They provide essential skills in designing rigorous research protocols, analyzing data from multiple sources, and determining the statistical significance of findings. Their input helps to ensure that conclusions are drawn based on solid scientific evidence and not on speculation or bias. This objective, data-driven approach is crucial in separating credible leads from anecdotal evidence or misinterpretations. The integration of rigorous statistical analysis is paramount in ensuring the reliability and credibility of the research findings.

Finally, the ethical considerations surrounding sea cryptid research necessitate the involvement of marine conservationists and environmental ethicists. Their expertise helps to ensure that investigations are conducted responsibly, minimizing the impact on marine ecosystems and the welfare of marine animals. The deep ocean is a fragile environment, and intrusive research methods could have unforeseen consequences. The collaborative input of conservationists ensures that research is conducted ethically and sustainably, prioritizing the well-being of both the marine environment and any potentially undiscovered creatures. The ethical framework is not simply an addendum; it is an integral part of the multidisciplinary approach, influencing research design and methodology at every step.

In conclusion, the future of sea cryptid research hinges on robust collaboration and interdisciplinary research. By uniting the expertise of cryptozoologists, marine biologists, oceanographers, geologists, engineers, geneticists, data analysts, and conservationists, we can harness a powerful synergy of skills and perspectives to conduct comprehensive, ethical, and scientifically rigorous investigations. This collaborative approach is not merely a matter of convenience; it is fundamental to the successful pursuit of these elusive creatures and to a deeper understanding of the mysteries that lie beneath the waves. The collective effort, combining varied expertise and integrating diverse methodologies, allows for a more complete picture of the underwater world, increasing the likelihood of both uncovering new species and promoting responsible exploration. This integrated approach is essential for navigating the complexities of the deep ocean, fostering a comprehensive and ethically responsible approach to research.

The success of any large-scale scientific endeavor often hinges on the participation of a wider community than just professional researchers. The vastness of the ocean and the inherent challenges of underwater exploration highlight the critical role of citizen science in sea cryptid research. Citizen science, involving the public in data collection and analysis, offers an unparalleled opportunity to expand the scope of investigations beyond the limitations of professional researchers' resources and expertise. This collaborative approach not only enhances data collection but also fosters public engagement and understanding of marine research and the cryptid investigation field.

One of the most significant contributions of citizen science lies in its potential to expand data collection exponentially. Professional researchers, despite their expertise, are limited by factors such as funding, time constraints, and geographical accessibility. Citizen scientists, on the other hand, can contribute data from a far wider geographical range, dramatically increasing the spatial coverage of research efforts. Imagine, for example, the effort required to monitor vast stretches of coastline for unusual marine phenomena. While a team of professional researchers might focus on a specific region, a network of citizen scientists distributed along the coast could provide a continuous stream of observations, significantly increasing the chances of detecting rare or unusual events. This distributed observation network can be particularly valuable for detecting transient phenomena or creatures with patchy distributions.

Furthermore, the diverse skills and interests of citizen scientists bring a unique perspective to data collection. While professional researchers might focus on specific parameters, citizen scientists can provide diverse observations, including anecdotal accounts, photographs, videos, and even acoustic recordings of unusual events. This multifaceted data collection enhances the comprehensiveness and reliability of research findings, enriching the understanding of potential cryptid behaviors and habitats. For instance, a seasoned fisherman might notice subtle behavioral patterns in a school of fish that a professional researcher might overlook, potentially indicating the presence of a predator not yet documented. Likewise, a keen-eyed birdwatcher might spot unusual surface activity that warrants further investigation.

The integration of citizen science requires careful planning and execution to ensure data quality and reliability. This includes providing citizen scientists with clear instructions and training on data collection methods, data recording protocols, and species identification guides. Establishing rigorous quality control measures is crucial to filter out erroneous or unreliable data. This might involve cross-referencing data from multiple sources, employing statistical methods to identify outliers, and verifying observations through independent verification processes. Such robust quality control is essential to maintaining the scientific integrity of the research.

The use of mobile applications and online platforms further facilitates citizen science participation. These technological tools allow for efficient data collection, storage, and sharing, enabling researchers to manage a large volume of data from numerous citizen scientists. Applications can incorporate features such as geographical tagging, time stamping, and photographic/video upload capabilities, allowing researchers to track observations accurately and assess their contextual relevance. Online platforms can serve as communication hubs, allowing researchers to communicate with citizen scientists, provide feedback, and disseminate research findings. The use of accessible technology significantly lowers the barrier to entry for participation, encouraging broader public involvement.

A successful citizen science program also requires effective communication and engagement with the public. Researchers should actively promote their projects through diverse channels, including social media, local media outlets, educational institutions, and community events. This outreach not only recruits citizen scientists but also raises public awareness of marine research and cryptid investigations, fostering a sense of community ownership and shared responsibility. Open communication builds trust and encourages participants to actively contribute and feel valued for their efforts.

Citizen science initiatives related to sea cryptid research can be structured around specific projects focused on particular geographic areas or target species. For example, a project might focus on collecting data on unusual marine sightings along a specific coastline, encouraging local residents to report sightings and provide accompanying photographic or video evidence. Another project might focus on analyzing underwater acoustic recordings for unusual sounds or vocalizations, recruiting citizen scientists to identify and classify these recordings. This targeted approach allows for more focused data collection and allows researchers to tailor training and quality control measures appropriately.

The ethical considerations surrounding citizen science participation must be carefully addressed. Protecting the privacy of citizen scientists, ensuring data security, and maintaining transparency in research methodologies are all crucial aspects of ethical engagement. Clear protocols outlining data ownership, usage rights, and attribution of findings should be established from the outset, ensuring the fair and respectful treatment of all participants. Furthermore, it's important to educate citizen scientists about responsible wildlife observation practices, emphasizing the importance of minimizing disturbance to marine ecosystems and wildlife. This includes guidelines on appropriate distances for observation, avoiding feeding or interacting with marine animals, and adhering to all relevant regulations and permits.

The benefits of citizen science in sea cryptid research extend beyond data collection. It plays a crucial role in raising public awareness and fostering a deeper understanding of marine environments and the importance of scientific investigation. By actively engaging the public, researchers can cultivate a sense of stewardship and responsibility towards the ocean. Citizen science also promotes scientific literacy, empowering individuals to become active participants in the scientific process. This collaborative approach fosters a sense of shared ownership and can inspire future generations of scientists and conservationists. The sense of community involvement fosters a deeper understanding and appreciation for both the ocean and the scientific method, leading to a more informed and engaged citizenry.

In conclusion, citizen science represents a powerful tool in advancing sea cryptid research. By leveraging the collective efforts and observations of the public, we can significantly expand the scope and depth of investigations, gaining access to data and insights that would otherwise be unattainable. However, the successful implementation of citizen science requires careful planning, effective communication, rigorous quality control, and a firm commitment to ethical practices. By fostering a collaborative and engaging environment, citizen science not only enhances the possibility of unlocking the mysteries of sea cryptids but also cultivates a broader understanding of, and responsibility toward, the ocean's vast and largely unexplored depths. The power of collective observation, combined with the rigorous approach of scientific methodology, offers a promising path towards unraveling some of the ocean's most enduring enigmas, while fostering a deeper connection between science and the public. The future of sea cryptid research, therefore, lies in the collaborative embrace of this powerful tool.

The pursuit of knowledge about elusive marine creatures, often termed sea cryptids, must always be tempered by a deep respect for the ocean's delicate ecosystems and the myriad life forms they support. Future research efforts, therefore, necessitate a robust ethical framework guiding every stage of investigation, from initial planning to the dissemination of findings. This framework must prioritize the preservation of marine habitats and the well-being of all marine life, ensuring that the quest for knowledge does not come at the expense of the environment or its inhabitants.

One critical ethical consideration revolves around the potential environmental impact of research activities. Underwater exploration, particularly using technologically advanced equipment like remotely operated vehicles (ROVs) and autonomous underwater vehicles (AUVs), must be conducted with utmost care to minimize disturbance to marine life and habitats. Carefully planned research expeditions should avoid sensitive areas like breeding grounds, migration routes, or fragile coral reef ecosystems. Detailed environmental impact assessments should be conducted before undertaking any significant research activity, identifying potential risks and implementing mitigation strategies to minimize disturbance. This might include restricting the duration and frequency of surveys in sensitive areas, adhering to strict operational protocols to avoid collision with marine life, and using environmentally friendly equipment and fuels.

The use of sonar and other acoustic technologies presents another important ethical challenge. While these tools are invaluable for underwater exploration, they can have detrimental effects on marine mammals, particularly whales and dolphins, who rely on sound for communication, navigation, and prey detection. High-intensity sonar can cause temporary or permanent hearing loss, disrupt communication patterns, and lead to behavioral changes. Therefore, responsible use of acoustic technologies requires adherence to strict guidelines and regulations, including limiting the intensity and duration of sonar pulses, employing frequency ranges that minimize impact on marine mammals, and incorporating real-time monitoring for the presence of marine mammals before and during sonar operations. Collaborating with marine mammal experts to establish appropriate mitigation strategies is crucial.

Similarly, the collection of biological samples, such as skin, tissue, or fecal matter, requires careful consideration. Any sampling activity should be conducted in a way that minimizes stress and harm to the animals. Non-invasive sampling techniques, when possible, should be prioritized. Researchers must obtain necessary permits and approvals from relevant authorities before collecting any biological samples, and all sampling activities should be guided by the principles of ethical animal handling and welfare. This includes adherence to best practices for minimizing stress and ensuring prompt and humane treatment of any captured animals.

The potential impact on indigenous communities and their cultural heritage must also be addressed. Many cultures have rich traditions and stories about sea monsters and other cryptids, often inextricably linked to their spiritual beliefs and livelihoods. Any research involving areas with indigenous populations requires respectful collaboration and engagement with the communities. This involves obtaining informed consent for research activities that may affect their lands or waters, respecting traditional knowledge and practices, and ensuring that research findings are shared appropriately with the communities. Avoiding appropriation or misrepresentation of indigenous knowledge is crucial, and ensuring fair and equitable benefit-sharing is essential.

The ethical implications extend beyond the environmental and social aspects. The dissemination of research findings also carries an ethical responsibility. Researchers must ensure that the information they share is accurate, transparent, and accessible. Any claims of cryptid sightings or evidence must be presented with caution and rigor, avoiding sensationalism and unsubstantiated speculation. The public must be clearly informed about the limitations of the evidence and the ongoing nature of the research. Open communication and transparency build public trust and encourage responsible engagement with the subject matter.

Furthermore, data security and privacy are crucial considerations. Any data collected during research expeditions, including photographic, video, and acoustic recordings, must be securely stored and protected from unauthorized access or misuse. Strict protocols for data management and security must be implemented to prevent data breaches and ensure the confidentiality of research information. Involving citizen scientists in data collection necessitates careful consideration of data privacy policies, ensuring that all contributions are handled responsibly and ethically.

The ethical responsibilities extend to the appropriate use of technology. Advanced technologies such as underwater drones and AI-powered image analysis tools offer significant potential for advancing sea cryptid research, but their use must be ethically sound. The use of AI, for example, raises concerns about bias in algorithms and the potential for misinterpretation of data. It is essential to use these tools responsibly, acknowledging their limitations and ensuring that their use does not lead to unfounded conclusions or misrepresentations of evidence.

Finally, the financial aspects of sea cryptid research must be ethically managed. Funding for research projects must be sourced and managed transparently, ensuring that funds are used effectively and efficiently to achieve the project's goals. Avoiding conflicts of interest and ensuring transparency in funding sources fosters public trust and confidence in the integrity of the research.

In conclusion, ethical considerations are paramount in the future of sea cryptid research. A robust ethical framework, encompassing environmental protection, social responsibility, data security, and transparent communication, is essential to ensure that the pursuit of knowledge does not compromise the integrity of the environment or the well-being of its inhabitants. By adhering to these principles, researchers can contribute to a more responsible and sustainable approach to exploring the mysteries of the deep, fostering a deeper understanding of both the ocean and our place within it. The future of sea cryptid research hinges not only on technological advancements but equally on our commitment to responsible and ethical practices. Only through such a conscientious approach can we hope to unravel the enigmas of the deep sea while preserving its wonders for generations to come.

The advancements in technology offer unprecedented opportunities to revolutionize sea cryptid research. The development of more sophisticated underwater vehicles, such as autonomous underwater vehicles (AUVs) and remotely operated vehicles (ROVs), equipped with high-resolution cameras and advanced sensors, allows for exploration of previously inaccessible depths and environments. These technologies enable researchers to gather high-quality visual data, acoustic data, and even collect biological samples from remote locations with minimal environmental impact. The increased maneuverability and endurance of these vehicles allow for extensive and repeated surveys of potentially cryptid-inhabited areas, significantly enhancing the chances of detecting elusive creatures.

Furthermore, the integration of artificial intelligence (AI) and machine learning (ML) into sea cryptid research holds immense potential. AI algorithms can analyze vast amounts of data collected from various sources, including sonar readings, underwater video footage, and eyewitness accounts, identifying patterns and anomalies that might be overlooked by human researchers. This automated analysis can dramatically accelerate the investigation process, highlighting potential cryptid sightings or unusual behaviors that warrant further investigation. ML algorithms can learn and adapt to new data, improving their accuracy and efficiency over time. For instance, AI could be trained to distinguish between known marine life and unusual forms, thus flagging potential cryptid candidates for closer examination by human experts.

The application of environmental DNA (eDNA) metabarcoding techniques promises to be a game-changer in sea cryptid research. This method involves extracting and analyzing DNA from environmental samples, such as water, sediment, or even seawater. By sequencing the DNA present in these samples, researchers can identify the species present in an area without needing to physically observe the organisms themselves. This technique is particularly useful for detecting cryptic or elusive species that are difficult to observe directly. The detection of unusual or unknown DNA sequences in environmental samples could indicate the presence of yet-undiscovered marine life, potentially including cryptids. This approach allows for a wider geographical scale of investigation, surpassing the limitations of traditional visual surveys.

Another promising technological avenue is the advancement in acoustic monitoring and analysis. Passive acoustic monitoring (PAM) systems can record underwater soundscapes over extended periods, detecting unusual sounds or vocalizations that could be attributed to unknown marine animals. The development of sophisticated algorithms for analyzing these acoustic data allows for the identification of unique sound signatures that might distinguish potential cryptids from known marine species. Furthermore, the integration of AI and ML in acoustic analysis can automate the process of identifying and classifying sounds, significantly speeding up the investigation. Combining PAM data with visual observations from underwater vehicles provides a multi-faceted approach, maximizing the chances of detecting and characterizing elusive creatures.

Satellite imagery and remote sensing techniques can also play a crucial role in predicting potential cryptid habitats and guiding research expeditions. Satellite data can provide information about oceanographic features such as water temperature, salinity, and currents, which can be used to identify areas suitable for certain types of marine life, including potential cryptids. By analyzing changes in oceanographic conditions over time, researchers can identify areas where unusual ecological events might be occurring, which could potentially indicate the presence of unknown creatures. The integration of satellite data with underwater surveys allows for a more comprehensive understanding of the cryptid's potential habitat and behavior.

The use of advanced underwater imaging techniques, including hyperspectral imaging and multispectral imaging, can provide crucial information about the morphology and characteristics of potential cryptids. These techniques capture images across a wider range of the electromagnetic spectrum than conventional cameras, revealing details about the color, texture, and composition of organisms that are invisible to the naked eye or standard cameras. This enhanced image quality can allow researchers to better distinguish cryptids from known species, particularly in challenging underwater environments.

Citizen science initiatives have emerged as invaluable tools in sea cryptid research, providing a vast network of potential observers and data collectors across the globe. The widespread availability of smartphones and underwater cameras allows for increased participation from the public, leading to a greater volume of sightings and observations. Well-designed citizen science projects, with clear guidelines and quality control measures, can contribute significantly to the data pool. The integration of online platforms and data management systems allows for the efficient compilation and analysis of citizen science data, further strengthening the reliability and scope of sea cryptid investigations. However, rigorous protocols for verifying observations and validating data remain crucial to maintain scientific integrity.

The future of sea cryptid research also necessitates increased collaboration between scientists from diverse disciplines. Integrating the expertise of marine biologists, oceanographers, physicists, engineers, computer scientists, and even anthropologists and historians is vital for a holistic approach. This interdisciplinary collaboration enables the development of innovative research methods, the synthesis of data from various sources, and the interpretation of findings within a broader scientific context. The sharing of knowledge and resources among researchers fosters a more efficient and productive research environment.

However, the ethical implications of these advanced technologies and approaches must be carefully considered. The deployment of sophisticated underwater vehicles and the use of AI algorithms raise concerns about potential environmental impacts, data privacy, and the responsible use of technology. Researchers must adopt a strict ethical framework, ensuring minimal disruption to marine ecosystems and maintaining transparency and accountability in their research practices. The potential for bias in AI algorithms and the need to avoid sensationalism in communicating research findings require careful attention.

In conclusion, the future of sea cryptid research appears bright, filled with the potential for groundbreaking discoveries. The convergence of technological advancements, innovative research methods, and increased interdisciplinary collaboration holds immense promise for unraveling the mysteries of the deep ocean. However, the pursuit of knowledge must be guided by a strong commitment to ethical principles, ensuring the responsible and sustainable exploration of our planet's oceans. The responsible application of these tools and techniques will be instrumental not only in addressing the enigma of sea cryptids, but also in deepening our understanding of the complex and often elusive biodiversity of the world's oceans, potentially unveiling new species and insights into marine ecosystems. The ethical considerations discussed earlier remain paramount; the pursuit of these fascinating creatures must always be secondary to protecting the very environment they inhabit. Only through this mindful approach can we hope to unlock the secrets of the deep sea while safeguarding its future for generations to come.

14

The Ongoing Quest

This investigation into the possibility of sea cryptids has taken us on a journey, traversing the murky waters of legend and folklore, while simultaneously navigating the rigorous currents of scientific inquiry. From the ancient mariners' tales whispered across centuries to the cutting-edge technologies of the modern era, we've examined a vast spectrum of evidence and methodologies. The overarching theme throughout this exploration has been the critical need for a balanced approach – one that embraces open-mindedness without sacrificing scientific rigor. We've learned that the pursuit of truth regarding these elusive creatures requires a delicate dance between imagination and skepticism, intuition and empirical evidence.

One of the most significant insights gained throughout this book lies in the recognition of the crucial role played by thorough, evidence-based research. Anecdotal evidence, while often the starting point for many cryptid investigations, can be highly subjective and prone to misinterpretation. Eyewitness accounts, often embellished by time or influenced by pre-existing biases, must be treated with caution and subjected to rigorous scrutiny. Similarly, photographic and video evidence requires careful analysis, considering factors such as lighting conditions, camera angles, and the potential for misidentification or manipulation. We've seen how seemingly compelling visual evidence can often be attributed to known marine life, optical illusions, or even naturally occurring phenomena.

The importance of establishing a robust baseline understanding of known marine life cannot be overstated. Before embarking on any sea cryptid investigation, investigators must develop a comprehensive knowledge of the biodiversity of the region under scrutiny. This includes familiarity with various species' appearances, behaviors, and habitats, enabling researchers to differentiate potential cryptids from known marine fauna. The exploration of this baseline knowledge frequently reveals instances where alleged cryptid sightings are, in fact, perfectly explainable observations of well-established species. We explored numerous examples where misidentifications stemming from limited knowledge have fueled misconceptions and exaggerated accounts.

Furthermore, the historical context of sea monster legends

proves crucial in navigating the complexities of sea cryptid research. By tracing the evolution of these narratives across different cultures and time periods, we gain valuable insights into the social, cultural, and environmental factors that might have shaped these beliefs. Analyzing historical accounts, alongside nautical charts and logs, can reveal patterns and recurring themes, offering clues to potential origins of these myths, and distinguishing between authentic observations and imaginative storytelling. This historical approach is not meant to dismiss anecdotal evidence entirely but rather to contextualize it, separating fact from fiction more effectively.

The book also underscored the importance of utilizing advanced investigative methodologies, embracing the technological advancements of our time. We have examined how sonar analysis, coupled with underwater drone operations and ROV deployments, can provide invaluable data for sea cryptid investigations. These technologies enable researchers to explore previously inaccessible depths and environments, providing access to habitats potentially occupied by elusive creatures. Acoustic data, analyzed using sophisticated algorithms, can detect unique sound signatures that might be attributed to unknown marine animals. The integration of AI and machine learning promises to revolutionize this field, facilitating the rapid analysis of vast datasets and identifying patterns that might escape human observation.

Equally important, we've discussed the transformative potential of eDNA metabarcoding, a technique that allows for the detection of species without needing direct observation. By analyzing environmental DNA from water or sediment samples, researchers can identify the presence of organisms, even cryptic or elusive ones, without ever seeing them. This method holds particular promise for sea cryptid research, offering a powerful tool for detecting the genetic signatures of unknown species within specific regions. It allows us to move beyond visual data, which is often limited by underwater conditions and the elusive nature of the potential cryptids.

However, the ethical considerations related to sea cryptid research must always remain at the forefront of any investigation. We must respect the sensitive marine ecosystems and prioritize their preservation. The deployment of advanced technologies must be conducted responsibly, minimizing environmental disturbance and adhering to strict ethical guidelines. The responsible use of AI algorithms, ensuring impartiality and transparency, is paramount. Misuse or poorly considered applications could lead to ecological damage or biased interpretations.

Throughout the book, a recurring emphasis has been placed

on the necessity of a collaborative approach. Sea cryptid research demands interdisciplinary collaboration, bringing together expertise from various scientific domains. The synthesis of data from marine biology, oceanography, physics, engineering, and computer science, among others, is vital for achieving a comprehensive understanding of these elusive creatures. The collective knowledge and resources contribute to more robust investigations and provide a broader perspective on the subject. Breaking down the silos between disciplines is paramount to progress in this field.

Moreover, the role of citizen science in sea cryptid research deserves considerable attention. The increasing accessibility of technology, in the form of underwater cameras and smartphones, coupled with the advent of online data sharing platforms, has empowered the public to contribute significantly to the collection of sightings and data. This participatory approach amplifies the scale of observation and increases the likelihood of identifying potentially significant events or data points. However, stringent quality control measures and data validation protocols are essential to maintain the scientific integrity of this citizen-sourced information.

In conclusion, while the evidence for the existence of sea cryptids remains inconclusive, the investigative methodologies outlined in this book offer a rigorous and ethical framework for exploring this fascinating area of research. The insights gained throughout this investigation highlight the importance of a balanced approach, combining imaginative exploration with unwavering scientific rigor. By embracing advanced technologies responsibly, collaborating across disciplines, and fostering a culture of open inquiry, we can continue to unravel the mysteries of the deep ocean, while simultaneously protecting and respecting the delicate balance of its ecosystems. The quest for these elusive creatures is not merely about proving or disproving their existence; it is about enhancing our understanding of the ocean's vast and largely unexplored depths, contributing to our knowledge of marine biodiversity, and ensuring responsible exploration for future generations. The ongoing investigation continues, but the path forward is now clearer, marked by a commitment to ethical, evidence-based research. The ocean's secrets are slowly being revealed, one carefully documented investigation at a time. The possibility of encountering the unexpected remains, and that alone fuels the continued pursuit of this enigma. The quest will continue, guided by the principles laid out here, until the mysteries are solved or we reach a conclusive understanding of what lies beneath the waves. The sea, in all its enigmatic glory, still holds its secrets close.

Despite the significant advancements in our understanding of marine life and the application of cutting-edge technologies, numerous unanswered questions persist regarding the possibility of sea cryptids. The very nature of these elusive creatures, their potential rarity, and the vastness of the ocean itself pose considerable challenges to definitive proof or disproof of their existence. Future research must address these challenges head-on, embracing novel approaches and fostering interdisciplinary collaboration to further illuminate this enigmatic field.

One critical area needing further investigation lies in refining our understanding of the limitations of current technologies. While sonar, underwater drones, and ROVs have revolutionized underwater exploration, these technologies are not without limitations. Sonar, for example, can be affected by factors such as water temperature, salinity, and the presence of marine sediment, potentially leading to false positives or missed detections. Underwater drones and ROVs, while offering unparalleled visual data, are often restricted by their operational range, battery life, and the challenges of navigating complex underwater environments. Further development of these technologies, potentially incorporating AI-driven obstacle avoidance and enhanced sensor capabilities, is essential. Miniaturization and increased autonomy of these platforms would greatly expand the scope of deep-sea exploration, enabling researchers to access previously unreachable habitats.

The development of novel technologies specific to sea cryptid research is also crucial. This includes the improvement of eDNA metabarcoding techniques. While eDNA offers a powerful non-invasive method for detecting unknown species, current methods might not be sensitive enough to detect extremely rare or elusive creatures. Advancements in DNA sequencing technology, coupled with the development of more sophisticated algorithms for analyzing eDNA datasets, are needed to increase the sensitivity and specificity of this approach. Furthermore, the development of new sensors capable of detecting unique bioluminescent signatures, acoustic signals, or even subtle chemical traces associated with unknown organisms would significantly enhance our ability to locate and identify potential cryptids. Exploration of advanced imaging techniques, capable of penetrating murky waters or overcoming issues of limited visibility, would also drastically improve observational capabilities.

Interdisciplinary collaboration will be paramount in advancing sea cryptid research. This involves integrating expertise from marine biology, oceanography, physics, engineering, computer science, and even fields like anthropology and history. Marine biologists can provide a baseline understanding of known marine species and their behaviors, while oceanographers can contribute expertise on ocean currents, water temperature profiles, and the distribution of various habitats. Physicists can help refine the accuracy of sonar and other acoustic technologies, while engineers can develop more advanced underwater robotics and sensors. Computer scientists can create advanced algorithms for analyzing large datasets and identifying subtle patterns that might indicate the presence of unknown organisms. Historians and anthropologists can provide insights into the cultural context of sea monster legends, helping to separate fact from fiction and identify potential clues.

A crucial element of future research involves a more formalized and systematic approach to data collection and analysis. Currently, much of the data regarding sea cryptid sightings is anecdotal, subjective, or lacks rigorous documentation. Implementing standardized protocols for collecting and documenting eyewitness accounts, photographic and video evidence, and sensor data is essential. This includes establishing clear guidelines for data quality control, ensuring the accuracy and reliability of information used in research. A centralized, globally accessible database, curated by a team of experts, could facilitate data sharing and collaboration among researchers, preventing duplication of effort and ensuring that data is analyzed consistently across various research projects. This collaborative database would be crucial in identifying potential patterns and correlations that might otherwise be missed when analyzing isolated incidents.

The ethical considerations of sea cryptid research cannot

be overemphasized. All investigations must prioritize the preservation of marine ecosystems and minimize any potential environmental impact. The use of advanced technologies must be carefully regulated and conducted responsibly, adhering to strict ethical guidelines. This includes obtaining the necessary permits for research activities, minimizing disturbances to marine life and habitats, and ensuring that any sampling or data collection does not harm the environment. Furthermore, a focus on education and public outreach is crucial to foster responsible citizen science and prevent actions that could compromise the integrity of marine environments. Ethical considerations must be central to all aspects of research, from the initial planning stages through the data collection and dissemination of findings.

The application of Artificial Intelligence (AI) and machine learning offers transformative potential for sea cryptid research. AI algorithms can be used to analyze vast quantities of data from various sources, identifying subtle patterns and anomalies that might escape human observation. AI can also be employed to enhance image and video analysis, improving the accuracy of species identification and detecting potential misidentifications. However, the use of AI in this field must be approached cautiously. It is crucial to ensure that AI algorithms are trained on unbiased datasets and that their outputs are carefully validated by human experts. Bias in training data could lead to skewed results and misinterpretations, potentially hindering rather than advancing the research. Transparency in the methodology, and rigorous testing and validation of AI algorithms are critical to maintain the scientific integrity of research outcomes.

Citizen science will continue to play a vital role in sea cryptid research. The involvement of the public in data collection can significantly expand the scope of observation, increasing the chances of detecting rare or elusive creatures. However, effective quality control mechanisms are necessary to maintain the scientific integrity of citizen-sourced data. This involves providing clear guidelines for data collection, implementing rigorous data validation procedures, and ensuring that citizen scientists receive adequate training and support. The establishment of online platforms for data submission and collaboration could effectively coordinate and manage this data flow, while promoting broader participation and engagement with ongoing research endeavors.

In conclusion, the quest for sea cryptids remains an ongoing endeavor, demanding a concerted effort from scientists, technologists, and the public alike. By addressing the unanswered questions, embracing new technologies responsibly, and fostering interdisciplinary collaboration, we can significantly advance our understanding of the ocean's mysteries. The future of sea cryptid research lies in a balanced approach, integrating rigorous scientific methodologies with imaginative exploration, always prioritizing ethical conduct and environmental stewardship. The ocean's depths hold secrets yet untold, and the persistent pursuit of knowledge, guided by these principles, offers the most promising path toward unraveling them. The journey continues, fueled by a commitment to scientific inquiry, technological innovation, and a profound respect for the intricate ecosystems that reside beneath the waves.

The pursuit of sea cryptids, while fueled by a sense of wonder and the thrill of discovery, necessitates a rigorous application of critical thinking and skepticism. Without these vital tools, the investigation risks veering into the realm of unsubstantiated claims and wishful thinking. Confirmation bias, the tendency to favor information that confirms pre-existing beliefs, is a significant pitfall to avoid. Researchers must actively combat this bias by consciously seeking out contradictory evidence and challenging their own assumptions. A truly objective investigation demands a willingness to consider alternative explanations, even if those explanations contradict initial hypotheses.

Consider the infamous "Loch Ness Monster" phenomenon. For decades, countless photographs, sonar readings, and eyewitness accounts have been presented as evidence for a large, unknown creature inhabiting the loch. However, a critical analysis of this evidence reveals a pattern of misinterpretations, hoaxes, and a lack of robust scientific validation. Many of the photographs, upon closer inspection, prove to be misidentified logs, boats, or other mundane objects. Sonar readings, while suggestive of large objects, often lack the resolution or context to definitively identify them. Eyewitness accounts, susceptible to memory distortion and suggestibility, frequently vary significantly in their descriptions. A critical approach requires carefully evaluating each piece of evidence, considering alternative explanations, and acknowledging the limitations of the available data. The absence of compelling, verifiable evidence, despite extensive investigation, should lead us to question the likelihood of a large, unknown creature residing within the Loch Ness. This exemplifies the importance of scrutinizing every claim with rigorous analysis rather than simply accepting evidence that confirms pre-existing beliefs.

Similarly, in investigations of other purported sea cryptids, it's crucial to differentiate between plausible hypotheses and extraordinary claims. While acknowledging the vast unexplored portions of the ocean, we must adhere to the principle of Occam's Razor – the simplest explanation is usually the best. Attributing unusual sightings to unknown creatures should only occur after meticulously ruling out more mundane possibilities, such as known marine life exhibiting unusual behavior, natural phenomena, or human-made artifacts. For instance, large aggregations of jellyfish or schools of fish can appear as large, indistinct shapes on sonar, leading to misinterpretations. Unusual underwater geological formations, such as deep-sea vents or methane seeps, can also create visual or acoustic anomalies. Thorough investigation necessitates a systematic approach, starting with the simplest, most readily testable hypotheses and only progressing to more extraordinary explanations when all other possibilities have been exhausted.

The maintenance of objectivity is paramount throughout the investigative process. Researchers must strive to remain impartial, avoiding emotional biases that might cloud their judgment. Personal beliefs, desires for discovery, or the pressure of public opinion should not influence the interpretation of data. Objectivity necessitates a transparent and rigorous methodology, where all data – including contradictory or negative findings – are documented and analyzed openly. Peer review, a cornerstone of scientific investigation, serves as a vital mechanism for ensuring objectivity. The scrutiny of experts in related fields helps identify potential biases, flaws in methodology, or misinterpretations of results. This process promotes rigorous evaluation and ensures that the conclusions drawn are based on sound evidence and careful analysis.

Furthermore, a critical approach demands an understanding of the limitations of technology and methodologies employed in the investigation. Modern technologies, such as underwater drones, sonar systems, and genetic analysis techniques, have revolutionized our ability to explore the ocean's depths. However, these technologies are not infallible. Sonar, for example, is susceptible to interference from various factors such as water temperature, salinity, and sediment, potentially leading to false positives or inaccuracies. Underwater imagery can be affected by low visibility conditions, making identification challenging. Genetic analysis, while powerful, requires careful sample collection and interpretation to avoid contamination or misidentification. A critical researcher acknowledges these limitations and incorporates them into their analysis, acknowledging the inherent uncertainties and biases associated with the tools they employ.

Moreover, critical thinking necessitates a keen awareness of the potential for deception and misrepresentation. The history of cryptozoology is rife with hoaxes and fabricated evidence. Researchers must possess the skills to detect fabricated evidence, recognize the hallmarks of deception, and evaluate the credibility of sources. This includes assessing the motivations of witnesses, considering the potential for exaggeration or misremembering, and evaluating the quality and authenticity of any photographic or video evidence presented. Digital manipulation of images and videos is increasingly prevalent, making it essential for investigators to possess the skills to authenticate such material and to differentiate between genuine evidence and sophisticated forgeries.

In addition to technological limitations and the potential for misrepresentation, cultural and historical context play a significant role in the interpretation of sea monster legends. Many sea monster stories have ancient origins, rooted in folklore and cultural beliefs. These stories, often passed down through generations, may reflect real-world observations of unusual marine life or natural phenomena that were subsequently embellished and transformed through oral tradition. Careful historical analysis can help differentiate between genuine observations and embellished narratives, separating fact from fiction. For example, historical accounts of sea monsters might describe large whales or squid, which are now well-understood, but which would have been unfamiliar to people in earlier eras.

Finally, a critical approach to sea cryptid investigations requires a commitment to ongoing learning and self-reflection. Researchers must remain open to new evidence and be willing to adjust their hypotheses or conclusions as new information emerges. This requires a scientific humility and a recognition of the limits of our current knowledge. The quest for sea cryptids is a process of continuous investigation and refinement, a journey driven by a dedication to truth-seeking rather than a preconceived notion of what should be found. It requires a balance of open-mindedness and critical skepticism, embracing the wonders of the unknown while simultaneously maintaining a rigorous and objective approach to its investigation. The ocean remains a vast and mysterious realm, and a balanced, scientifically-grounded approach is essential in our quest to unravel its secrets.

The preceding discussion highlighted the crucial role of critical thinking and skepticism in navigating the often-murky waters of cryptozoological investigation. However, a balanced approach demands not only rigorous scrutiny but also a willingness to embrace the unknown, to consider possibilities that initially seem improbable. The pursuit of truth, particularly in a field as enigmatic as marine cryptozoology, necessitates a delicate dance between healthy skepticism and open-mindedness. A purely skeptical approach, rigidly dismissing all unusual observations without thorough investigation, risks overlooking genuine anomalies. Conversely, unbridled enthusiasm, devoid of critical analysis, can easily lead to misinterpretations and the propagation of misinformation.

The ocean, a vast and largely unexplored environment, harbors countless mysteries. Our current understanding of marine life, while extensive, is undeniably incomplete. The sheer size and complexity of the oceanic ecosystem, coupled with technological limitations, means that much remains unknown, potentially including undiscovered species, unusual behaviors, and phenomena that defy easy explanation. To dismiss all unexplained marine phenomena as illusions, hoaxes, or misidentifications would be a significant oversight, potentially hindering the discovery of genuine scientific breakthroughs.

Consider, for instance, the ongoing debate surrounding giant squid. For centuries, these creatures existed primarily in the realm of myth and legend, dismissed as mere folklore. Yet, persistent accounts, coupled with the occasional discovery of colossal squid specimens, have definitively proven their existence. The initial skepticism surrounding these reports, while understandable given the lack of substantial evidence, ultimately hampered early investigations. A more open-minded approach, coupled with advancements in underwater exploration technology, has significantly advanced our understanding of this enigmatic creature. This example highlights the importance of maintaining a receptive posture towards unusual reports, while simultaneously applying the most rigorous scientific methods to verify their validity.

Similarly, the investigation of bioluminescence in the deep ocean exemplifies the value of open-mindedness. The extraordinary displays of light produced by various marine organisms, once thought to be fanciful exaggerations, are now well-documented and understood as essential aspects of deep-sea ecosystems. Initial skepticism towards such reports, stemming from a lack of visual evidence and a limited understanding of deep-sea environments, delayed scientific exploration in this area. Had researchers remained rigidly skeptical, discarding all reports of light displays as unverifiable anomalies, we would have missed out on understanding a fundamental aspect of marine biology. Thus, a willingness to accept the possibility of the unexpected, combined with rigorous investigation, is crucial in advancing scientific knowledge.

Furthermore, the pursuit of sea cryptids often necessitates interdisciplinary collaboration. Effective investigations often require expertise in marine biology, oceanography, geology, history, and even computer science and image analysis. Bringing together experts from various fields allows for a more holistic approach, providing different perspectives and methodologies. Open-mindedness is essential in this collaborative context, as it fosters respectful dialogue and prevents the dismissal of valuable insights based on preconceived notions or disciplinary biases. Each discipline brings its unique set of tools and perspectives; a truly comprehensive investigation benefits from the contributions of all.

The ethical considerations involved in sea cryptid research further emphasize the need for an open and responsible approach. Conducting research in sensitive marine ecosystems necessitates a commitment to environmental protection and the preservation of biodiversity. This requires carefully considering the potential impact of any research activities on the environment, adopting non-invasive techniques wherever possible and strictly adhering to all relevant regulations. An open and collaborative approach is essential to ensure transparency and accountability in such research efforts, thus maintaining the integrity of both the investigation and the ecosystem under scrutiny.

The integration of historical accounts and folklore into modern investigations should not be dismissed outright. While recognizing the limitations of oral histories and anecdotal evidence, a careful examination of historical records can often provide valuable contextual information. Sea monster legends, passed down through generations, sometimes contain surprisingly accurate observations of unusual marine life or events. Dismissing these narratives entirely, without acknowledging their potential to provide clues or insights, may lead to the overlooking of crucial information. A critical approach would involve examining these accounts for verifiable elements, cross-referencing them with other historical records and scientific data, and separating demonstrable facts from embellishment or myth.

Equally important is the need for transparency and reproducibility. The scientific method relies on the ability of other researchers to replicate studies and verify results. This means that all data, methods, and analysis must be clearly documented and made available to the broader scientific community. Open-mindedness in this context involves a willingness to share data, even negative findings, which are often overlooked but equally crucial for ensuring the robustness of conclusions drawn. Transparency promotes credibility, allowing for rigorous peer review and ultimately strengthening the validity of any findings.

Finally, it is crucial to acknowledge the inherent uncertainties involved in exploring the unknown. The quest for sea cryptids is not a search for definitive answers but rather a journey of discovery, a continuous process of refinement and learning. Open-mindedness in this context means accepting that there might be answers beyond our current understanding, and that many questions may remain unanswered for a considerable amount of time. It is a balance of seeking clarity and acknowledging the limits of human knowledge. This humility is paramount. While rigorous analysis and skepticism are necessary to filter out unsubstantiated claims, a closed-minded approach could limit the scope of discovery, preventing progress and hindering the advancement of scientific knowledge. The ocean's depths remain largely unexplored, and a blend of rigorous skepticism and open-minded curiosity is the best approach to unraveling the remaining mysteries of the deep.

The preceding discussion underscores the critical interplay between rigorous skepticism and open-mindedness in navigating the complexities of marine cryptozoology. However, the journey towards understanding the ocean's enigmas requires more than individual effort; it necessitates a collective, collaborative approach. The field's future hinges on the active participation of a new generation of researchers, enthusiasts, and citizen scientists, all working together to systematically explore the unexplored.

This call to action is not merely an invitation to join the hunt for elusive creatures; it's an appeal to embrace a new paradigm of scientific inquiry—one characterized by interdisciplinary collaboration, technological innovation, and a commitment to ethical research practices. The vastness of the ocean, coupled with the limitations of current technology, necessitates a multifaceted approach. No single individual, or even single discipline, possesses the entire toolkit needed to tackle the challenges presented by cryptozoology.

The first and perhaps most crucial step is fostering interdisciplinary collaboration. Marine cryptozoology is inherently multi-faceted. Successful investigations require expertise not only in marine biology but also in oceanography, geology, history, computer science, image analysis, and even linguistics (for deciphering historical accounts and folklore). Each discipline brings a unique perspective and a set of specialized tools. For example, a marine biologist might identify potential habitats based on known species distribution, while an oceanographer could analyze currents and water temperatures to pinpoint areas of unusual activity. A historian could shed light on the historical context of sightings, while a computer scientist could develop sophisticated algorithms for analyzing sonar data or satellite imagery. The synergy created by this diverse team is far greater than the sum of its individual parts.

Furthermore, technological advancements are vital to furthering our understanding of the oceans' mysteries. Improvements in underwater drone technology, remotely operated vehicles (ROVs), and autonomous underwater vehicles (AUVs) have already revolutionized our ability to explore the deep sea. These advancements allow for more extensive and detailed surveys of previously inaccessible habitats, increasing the chances of encountering unusual or unknown species. However, technology alone is insufficient. The effective deployment and analysis of data gathered from these sophisticated tools requires skilled operators and data scientists. Therefore, the call to action also encompasses the need for training and education in the operation and analysis of advanced marine research technologies.

Beyond technology, the ethical considerations inherent in marine cryptozoology cannot be overlooked. Researchers must prioritize the preservation of marine ecosystems and the well-being of all marine life. Non-invasive research techniques should always be prioritized, with a focus on minimizing the disturbance of sensitive habitats. This requires adherence to stringent ethical guidelines and collaboration with relevant regulatory bodies. Transparency in research methods and data sharing is also crucial for maintaining the integrity of the field. The open sharing of data, both positive and negative findings, encourages scrutiny, promotes reproducibility, and safeguards against the propagation of misinformation.

In addition to technological innovation and ethical considerations, the integration of historical accounts and folklore into modern investigations remains vital. While acknowledging the limitations of anecdotal evidence, it's imperative to consider the potential value of historical records. Ancient seafaring tales and legends, passed down through generations, often contain remarkably accurate descriptions of unusual marine life or phenomena. These historical accounts, carefully analyzed and cross-referenced with scientific data, may provide crucial contextual information and guide modern investigations towards promising research avenues. This necessitates expertise in historical research methods, linguistic analysis (especially in cases of older texts or languages), and the ability to discern credible information from embellishment and myth.

Another critical aspect of this call to action is the fostering of citizen science initiatives. The vastness of the oceans makes comprehensive scientific surveys prohibitively expensive and time-consuming. Citizen science projects, involving volunteer observers and data collectors, can significantly expand the scope of research. These projects require careful planning and management to ensure data quality and consistency, but the potential benefits are immense. By engaging the public in the process of scientific inquiry, these initiatives can not only generate valuable data but also raise awareness about the importance of marine conservation and foster a sense of shared responsibility in protecting our oceans.

Finally, this call to action extends beyond professional researchers. Enthusiasts, amateur naturalists, and divers can make significant contributions through meticulous observation, careful documentation, and responsible reporting of any unusual marine sightings. Their observations, while requiring careful scrutiny and verification, can offer valuable clues and guide more focused investigations. The key lies in responsible reporting—providing accurate details of location, time, and circumstances, supporting the observation with photographs or videos whenever possible, and understanding the importance of maintaining data integrity.

In conclusion, the pursuit of marine cryptozoology demands a collective effort. The exploration of the ocean's unexplored depths requires a multidisciplinary approach, leveraging technological advancements while upholding the highest ethical standards. By fostering interdisciplinary collaboration, integrating historical insights, embracing citizen science initiatives, and encouraging responsible reporting from enthusiasts, we can significantly accelerate the pace of discovery and deepen our understanding of the ocean's mysteries. The future of marine cryptozoology rests not on individual efforts alone, but on a vibrant community united by a shared passion for exploration, scientific rigor, and a commitment to unveiling the truth hidden within the ocean's depths. The potential discoveries—and the deeper understanding of our planet's complex ecosystems—await those who answer this call. The ocean holds its secrets close, but with collaborative effort, dedication, and unwavering commitment to ethical scientific exploration, these mysteries are within our reach. The time for action is now.

15

Appendix: Resources and Further Reading

The preceding discussion highlights the crucial need for a collaborative, multidisciplinary approach to marine cryptozoology. To facilitate this collaboration, access to relevant resources and the network of established organizations and institutions is paramount. This section provides a curated list of organizations and institutions actively involved in marine biology, oceanography, cryptozoology, and related fields that can serve as invaluable resources for researchers, enthusiasts, and anyone interested in learning more about the subject. It is important to note that this list is not exhaustive, and many other organizations contribute to the field; however, these represent a strong starting point for further exploration.

We begin with organizations focused on marine biology and oceanography, providing the foundational scientific knowledge crucial for evaluating potential cryptid sightings and differentiating them from known species. The **Scripps Institution of Oceanography** at the University of California, San Diego, stands out as a world-renowned center for oceanographic research. Their extensive database of marine life, coupled with their ongoing research projects on deep-sea ecosystems, makes them an invaluable resource. Their website offers access to publications, data sets, and educational materials.

Similarly, the **Woods Hole Oceanographic Institution** (WHOI) on Cape Cod, Massachusetts, is another leading institution with a rich history of groundbreaking discoveries in marine science. WHOI's research spans a wide range of disciplines, including marine biology, ocean engineering, and geological oceanography, providing diverse perspectives relevant to the study of marine cryptids. Their online resources include research publications, educational programs, and information on ongoing expeditions.

Moving beyond the strictly scientific realm, organizations focused on marine conservation and environmental protection play a crucial role. Their work is fundamentally intertwined with cryptozoology, as the protection and study of unknown marine life requires a keen awareness of the impact of human activity on marine ecosystems. The **World Wildlife Fund** (WWF) and the **Ocean Conservancy** are two prominent organizations that actively work towards conserving marine biodiversity and addressing threats such as pollution and habitat destruction. These organizations not only contribute to the overall body of knowledge regarding marine life, but they also provide a valuable framework for conducting ethical and responsible research. Their websites offer access to information on marine conservation efforts, environmental policy, and advocacy initiatives.

The field of cryptozoology, though often perceived as fringe, has its own network of researchers, societies, and institutions. The **International Cryptozoology Museum** located in Portland, Maine, houses a unique collection of artifacts and archival material related to the study of cryptids. Their website offers a window into the history of cryptozoological investigations and provides a platform for enthusiasts to learn about the field. While the museum itself doesn't conduct active fieldwork, it serves as an invaluable repository of information and a center for discussion and critical evaluation of cryptozoological claims. It's crucial to approach such organizations with a critical eye, considering the varying levels of scientific rigor applied to their investigations. The museum's role lies in preserving and showcasing the cultural and historical significance of cryptozoology, providing context for modern investigations.

In addition to specialized organizations, numerous universities and research institutions around the world contribute to the study of marine life and related fields. Departments of marine biology, oceanography, and zoology in major universities often have researchers involved in projects that indirectly inform the understanding of marine cryptozoology. For instance, the **University of Hawaii**'s School of Ocean and Earth Science and Technology is known for its extensive work on marine biodiversity in the Pacific Ocean. Similarly, the **University of Oxford**'s Department of Zoology, with its long-standing history of research, contributes to the wider scientific understanding of animal behaviour and evolution, offering valuable comparative perspectives for cryptozoological inquiries. It's advisable to search the online databases of specific university departments to discover ongoing research projects that might be relevant to marine cryptozoology.

Furthermore, government agencies and research organizations play a crucial role. Agencies such as the **National Oceanic and Atmospheric Administration** (NOAA) in the United States conduct extensive research on the oceans, producing a wealth of data on oceanographic conditions, marine species distribution, and underwater environments. Their extensive data sets, often publicly accessible, can be invaluable in evaluating cryptozoological claims and contextualizing potential sightings. Similarly, other national and international government agencies with marine research components are vital resources, often holding extensive datasets. Their research reports and publications can provide baseline information on marine fauna and habitats. Access to these datasets may require navigating specific agency websites and understanding their data access policies.

The realm of technological advancements further expands the network of relevant resources. Companies specializing in underwater technology, such as those producing remotely operated vehicles (ROVs) and autonomous underwater vehicles (AUVs), contribute indirectly but significantly to the field. Their technological innovations allow for deeper and more comprehensive exploration of marine environments, opening up previously inaccessible areas to investigation. However, direct access to these technologies is usually limited to researchers and specialized institutions.

Finally, online resources play a significant role. Several databases, journals, and online forums dedicated to marine biology, oceanography, and cryptozoology offer valuable information and discussion platforms. While evaluating information found online requires careful scrutiny and critical thinking (given the variable quality of online sources), online databases like those maintained by NOAA and scientific journals like *Marine Biology* and *Deep Sea Research* offer validated information. Forums and online communities can provide opportunities for discussions, collaboration, and the sharing of observations. However, it's essential to be aware of potential biases and misinformation circulating online. Cross-referencing information from multiple sources and evaluating the credibility of the source are crucial to ensuring the reliability of the information gathered online.

In conclusion, navigating the world of marine cryptozoology necessitates the utilization of a broad array of resources. This list of organizations and institutions serves as a starting point for researchers and enthusiasts seeking to contribute to this fascinating and challenging field. Remember that the successful investigation of marine cryptids requires a combination of scientific rigor, ethical considerations, interdisciplinary collaboration, and access to a wide range of resources. The organisations and institutions listed represent only a small fraction of the vast network of individuals and entities contributing to the understanding of our oceans and their mysteries. Further research into specialized fields and the use of online databases will help reveal many other crucial resources for the aspiring marine cryptozoologist. The journey of discovery demands dedication, persistence, and a willingness to engage with a multifaceted array of resources. The pursuit of the ocean's enigmatic inhabitants is a shared endeavor, reliant on collaboration and the collective knowledge of a diverse scientific community.

This section provides a curated list of recommended books and articles to deepen your understanding of marine biology, cryptozoology, and related fields. This reading list is designed to complement the practical investigative techniques discussed earlier and offers a range of perspectives, from established scientific literature to explorations into the more speculative realm of cryptozoology. Remember that critical thinking and careful evaluation of sources are crucial when navigating this diverse body of work. Not all sources hold equal weight in terms of scientific rigor; some provide intriguing narratives and historical context, while others offer robust scientific data. It is important to approach each text with a discerning eye, considering the author's background, methodology, and the overall strength of their evidence.

We begin with foundational texts in marine biology, providing the necessary background for distinguishing potential cryptids from known species. "Marine Biology" by Peter Castro and Michael Huber is a comprehensive textbook often used in undergraduate programs. Its breadth of coverage, encompassing various aspects of marine ecosystems and organisms, provides an excellent foundation for understanding the biodiversity of the oceans. For a more specialized focus on deep-sea biology, "Deep-Sea Biology: A Natural History of Organisms at the Deep Seafloor" by Robert Carney provides invaluable insights into the unique organisms and habitats found in the abyssal zone, where many cryptids are purported to reside. Understanding the adaptations and lifestyles of deep-sea creatures is crucial for evaluating the plausibility of deep-sea cryptid sightings. These texts offer detailed anatomical descriptions, ecological relationships, and evolutionary histories, providing a baseline for comparison when evaluating unusual sightings.

Moving into the realm of oceanography, "Oceanography: An Invitation to Marine Science" by Harold Thurman and Alan Trujillo offers a broad overview of oceanographic processes, currents, and the physical and chemical properties of the marine environment. Understanding these aspects is fundamental to understanding the distribution of marine life, including potential cryptids. The book's focus on oceanographic data collection and analysis also provides invaluable context for interpreting evidence gathered during cryptid investigations. For a more in-depth study of specific oceanographic phenomena relevant to cryptid investigations, such as unusual currents or deep-sea geological formations, specialized literature may be necessary. Searching scientific databases like Web of Science or Scopus using keywords such as "deep-sea currents," "oceanographic anomalies," or "hydrothermal vents" will lead you to numerous research articles published in journals like *Deep Sea Research Part I: Oceanographic Research Papers* and *Journal of Geophysical Research: Oceans.* Careful examination of these peer-reviewed publications can provide crucial scientific context for assessing the plausibility of potential cryptid sightings.

Further expanding our scope to include related fields, "Biological Oceanography: An Introduction" by Alan Longhurst explores the dynamic interplay between marine organisms and their physical environment. This understanding is crucial for evaluating the ecological context of potential cryptid sightings. Understanding the distribution patterns of known species, their feeding habits, and their interactions with their environment helps determine whether a purported cryptid sighting fits within the known ecological framework. Conversely, discrepancies between an observed creature and known species may suggest something more unusual, yet careful investigation is essential to avoid misinterpretations.

Now, we delve into the realm of cryptozoology, a field requiring a critical and discerning approach. The book "Cryptozoology A to Z: The Encyclopedia of Loch Ness Monsters, Bigfoot, Chupacabra and Other Enigma" by Loren Coleman provides an encyclopedic overview of various cryptids, including those from the marine environment. While it approaches the subject with an openness to possibility, it's vital to approach Coleman's work, as well as other cryptozoological literature, with a critical eye. Assess the quality of evidence presented, evaluate the methodologies employed, and consider alternative explanations for the reported phenomena.

While many books on cryptozoology offer speculative narratives, focusing on robust scientific methodology remains paramount. Therefore, it is essential to supplement any reading in cryptozoology with rigorous research in marine biology and oceanography. The balance lies in maintaining an open mind to possibilities while rigorously evaluating evidence within a scientific framework. In this regard, combining research on specific marine animals and habitats with explorations of historical accounts of sea monsters helps to form a comprehensive approach. For example, studying the behavior and appearance of giant squid can be supplemented by historical accounts of colossal squid-like creatures from sailors' logs. Such a multifaceted approach allows for a nuanced investigation, integrating scientific data with historical context.

Numerous articles can further enrich your understanding. Searching academic databases like JSTOR and Google Scholar with keywords such as "marine cryptids," "unidentified marine organisms," "historical accounts of sea monsters," and similar terms will yield a range of articles focusing on specific cryptid cases, analysis of eyewitness accounts, and discussions on the methodology of cryptozoological research. Remember to prioritize peer-reviewed articles published in reputable scientific journals, as they generally undergo rigorous scrutiny. Many scientific publications may indirectly contribute to the field by examining specific marine phenomena, such as unusual deep-sea bioluminescence or unusual acoustic signals that could potentially be associated with unknown marine organisms.

Furthermore, examining historical documents, such as ship logs, nautical charts, and early natural history writings, can provide valuable insights into past encounters with potential cryptids. Digitizing projects that make such documents accessible online are invaluable resources. Libraries specializing in maritime history or archival centers can also contain pertinent information. However, be aware that historical accounts are often subject to biases and embellishments, and careful critical analysis is necessary to separate factual observations from folklore or exaggeration.

Finally, engaging with online communities and forums dedicated to marine biology and cryptozoology can facilitate collaborative discussions and sharing of information. However, it's crucial to approach these resources critically, evaluating the credibility of sources and discerning fact from speculation. Active participation in such communities can, however, provide valuable networking opportunities and expose you to alternative viewpoints.

In conclusion, the resources presented offer a starting point for a deeper dive into the fascinating world of marine cryptozoology. Remember that the investigation of marine cryptids requires a blend of scientific rigor, historical perspective, and a critical approach to information. The combination of established scientific literature, explorations into the speculative, and a rigorous, analytical mindset will equip you to navigate the complexities of this field effectively. The exploration of the ocean's mysteries is a continuous journey requiring a constant process of learning and refinement. This reading list serves as a guidepost along that path, urging you to explore, question, and contribute to the expanding knowledge of our oceans and their enigmatic inhabitants.

This section focuses on invaluable online resources that can significantly aid your investigations into marine cryptids. The digital age offers unprecedented access to information, and effectively utilizing online resources is crucial for any modern cryptozoological investigation. This is not merely about finding sensationalized accounts; rather, it's about accessing credible data, scientific literature, historical records, and collaborative platforms. Remember, critical evaluation remains paramount, and you should always cross-reference information from multiple sources to ensure accuracy and reliability.

We begin with major scientific databases. The most prominent are undoubtedly Web of Science and Scopus, two comprehensive repositories of peer-reviewed scientific literature. These databases allow you to search for articles, dissertations, and other scholarly works related to marine biology, oceanography, and related fields. Use precise keywords such as "unidentified marine organisms," "bioluminescent deep-sea fauna," "unusual acoustic signals in the ocean," or specific cryptid names (though always acknowledging the speculative nature of these terms within a scientific context). Refine your searches using filters for publication date, journal impact factor, and other relevant criteria. This ensures you're focusing on high-quality, credible research. Don't underestimate the power of Boolean operators (AND, OR, NOT) to refine your searches and extract the most relevant information. For example, a search for ("giant squid" AND "behavior" AND "deep sea") will yield articles specifically focusing on giant squid behavior in the deep sea, a crucial aspect of understanding potential misidentification of deep-sea cryptids.

Beyond Web of Science and Scopus, other databases offer valuable insights. JSTOR provides access to a vast archive of academic journals, books, and primary source materials, including historical accounts of marine life observations. Many universities and research institutions also maintain online repositories of their research data, which can include raw data from oceanographic surveys, sonar scans, or underwater video footage – potentially containing unforeseen clues relevant to your investigations. Searching these institutional repositories can unearth valuable data that might not be readily available through commercial databases. Remember to always respect data usage policies and obtain necessary permissions before using any data for your research.

Image archives play a vital role. Sites like Flickr, Wikimedia Commons, and even commercial stock photo databases may contain relevant imagery, such as historical photographs of unusual marine life sightings, or images captured by underwater exploration vessels. However, be extremely cautious when assessing the authenticity and provenance of images. Many images online are mislabeled, digitally manipulated, or outright hoaxes. Always examine the metadata accompanying images (date, location, source) and critically evaluate the context within which the image was obtained. Cross-referencing images with other sources, such as historical accounts or scientific literature, can help you establish their reliability. Specialized databases held by museums and research institutions, often accessible through their websites, provide higher quality and more reliably sourced images.

Government agencies also offer valuable online resources. The National Oceanic and Atmospheric Administration (NOAA) in the United States, along with equivalent agencies in other countries, maintain extensive websites with data on oceanographic research, marine life surveys, and environmental monitoring. Their websites often include data visualizations, maps, and downloadable datasets that can contribute significantly to your research. Their data may not directly relate to cryptids, but can provide crucial context about the environments where cryptids are reported to exist. For example, NOAA data on deep-sea currents and hydrothermal vents can help you determine the plausibility of certain cryptid sightings, by clarifying whether known species are likely to populate an area, or if there might be unexplored environmental factors that could support unfamiliar organisms.

Moving beyond formal scientific resources, online forums and communities focused on marine biology and cryptozoology can offer a space for discussion and collaboration. Sites like Reddit (subreddits such as r/MarineBiology and, with caution, relevant cryptozoology subreddits), online forums dedicated to specific cryptids, or even social media groups focused on ocean exploration can provide access to a wealth of information, including firsthand accounts, eyewitness testimonies, and diverse interpretations of evidence. However, remember that information obtained from these sources must be carefully vetted for accuracy and reliability. The anonymity afforded by online forums can lead to a proliferation of unsubstantiated claims and misinformation. Therefore, always treat information from these sources with a high degree of skepticism, confirming details through other channels whenever possible.

Mapping tools are particularly useful. Online mapping services like Google Earth and ArcGIS can assist in the geolocation of sightings, allowing you to analyze the geographical distribution of reported events. Mapping the locations of cryptid sightings, along with environmental data such as water depth, temperature, and salinity, can reveal potential patterns or correlations that might shed light on the nature of these elusive creatures.

Finally, digital archives of historical documents are invaluable. Many libraries and archives have digitized historical ship logs, nautical charts, and natural history writings, making them easily accessible online. These primary sources can provide crucial historical context for sea monster legends and reported sightings. However, remember that historical documents are often subject to bias, exaggeration, and misinterpretation. Critically evaluate the source, the author's potential biases, and the historical context in which the account was created. Compare multiple accounts whenever possible to identify recurring themes or discrepancies.

In conclusion, the online world offers a vast and ever-expanding repository of information relevant to the investigation of marine cryptids. However, a discerning and critical approach is paramount. Always verify information from multiple sources, prioritize peer-reviewed scientific literature, evaluate the credibility of online communities, and thoroughly examine the context in which information is presented. By effectively utilizing these digital resources, coupled with sound investigative methodologies, you can significantly enhance your capacity to explore the mysteries of our oceans and potentially unveil the truth behind sea monster legends. Remember, the responsible and ethical investigation of potential cryptids should always prioritize the protection of marine ecosystems and the avoidance of misinformation.

This glossary provides definitions for key terms used throughout "An Investigator's Guide to Sea Cryptids." Understanding these terms is crucial for effective communication and analysis within the field of cryptozoology, and for navigating the often complex language of marine biology and investigative methodologies. The definitions provided here aim to be clear, concise, and accessible to a broad readership, from seasoned researchers to curious amateurs.

We will cover terms relating to marine biology, cryptozoology, investigative techniques, and data analysis, providing a comprehensive reference for your investigations.

Marine Biology Terms:

- **Benthic Zone:** The lowest level of a body of water, including the sediment surface and some subsurface layers. This zone is home to a vast array of organisms adapted to life on or within the seabed. Understanding the benthic zone is vital when investigating deep-sea cryptids, as many potential cryptids are hypothesized to inhabit this environment. Differences in benthic habitat, such as hydrothermal vents or cold seeps, will also greatly influence the types of organisms found there.
- **Bioluminescence:** The production and emission of light by a living organism. Many deep-sea creatures utilize bioluminescence for various purposes, including attracting prey, deterring predators, or communication. The observation of unusual bioluminescent patterns in the ocean could be indicative of unknown species. Detailed study of bioluminescent patterns, including their spectral characteristics and temporal variations, can contribute significantly to cryptid investigations.
- **Cephalopod:** A class of mollusks characterized by their bilateral body symmetry, a prominent head, and tentacles. This class includes squids, octopuses, cuttle-

fish, and nautiluses. Giant squid and colossal squid are often invoked in discussions of sea monsters, making understanding cephalopod biology crucial. Their diverse adaptations to deep-sea environments highlight the complexity of the marine ecosystem and the possibility of undiscovered species.

- **Deep Scattering Layer (DSL):** A layer in the ocean where sound waves are strongly scattered, often due to the presence of large numbers of marine organisms. The DSL's composition varies depending on location and time of day, but it often includes a mixture of fish, crustaceans, and other organisms, making it a potential source of misidentification in sonar data. Careful analysis of sonar returns, alongside environmental factors such as depth and temperature, is essential for differentiating between known DSL constituents and potential cryptids.
- **Hydrothermal Vent:** A fissure in a planet's surface from which geothermally heated water issues. These vents are found on the ocean floor, and are home to unique ecosystems supported by chemosynthesis rather than photosynthesis. Hydrothermal vents are a prime area for the discovery of novel species, as the extreme conditions create unique ecological niches. Investigating the biodiversity of these ecosystems is essential for establishing a comprehensive understanding of deep-sea life and determining the likelihood of undiscovered species.

- **Nekton:** Actively swimming aquatic animals that can move independently of water currents. Nekton includes fish, marine mammals, cephalopods, and sea turtles. Many potential cryptids are hypothesized to be nektonic, meaning their movement patterns and behaviors need to be carefully considered when analyzing sightings. Understanding the swimming capabilities and migratory patterns of various nektonic species can help to differentiate between known species and potential cryptids.
- **Pelagic Zone:** The open ocean water column, excluding the benthic zone. The pelagic zone is vast and largely unexplored, making it a prime habitat for undiscovered species. The pelagic zone can be further subdivided into different zones based on depth, each possessing distinct characteristics and supporting unique communities of organisms. Understanding these zones and their inhabitants is crucial for assessing the plausibility of cryptid sightings.
- **Plankton:** Organisms drifting in water currents. Plankton comprises phytoplankton (plants) and zooplankton (animals), forming the base of many marine food webs. While individually small, plankton aggregations can create significant sonar reflections, which need to be accounted for when interpreting sonar data. Misidentification of large plankton aggregations as unknown creatures is a common pitfall.
- **Zooplankton:** Animal plankton, including various microscopic and macroscopic organisms such as

copepods, krill, jellyfish, and larval forms of other marine creatures. Zooplankton's significant role in the marine ecosystem, and its abundance in certain areas, needs careful consideration in investigations. Unusual aggregations of zooplankton, or the discovery of unusually large zooplankton, could be indicative of environmental changes or the presence of unknown species.

Cryptozoology Terms:

- **Cryptozoology:** The study of hidden or undiscovered animals. This field involves investigating reports of animals not yet scientifically recognized or documented. Cryptozoology employs a range of investigative techniques, from analyzing eyewitness accounts and physical evidence to utilizing advanced technologies like sonar and underwater drones. It's crucial in cryptozoological investigations to maintain a balance between open-mindedness and scientific rigor, acknowledging the speculative nature of many claims.
- **Eyewitness Testimony:** Accounts of events or sightings provided by individuals who have witnessed them. Eyewitness testimony is an important component of cryptozoological investigations, but it must be carefully evaluated and cross-referenced with other evidence to account for potential biases, inaccuracies, or misinterpretations. It's vital to consider factors like lighting conditions, distance, the observer's prior

knowledge, and potential psychological factors that could influence the reliability of an eyewitness account.

- **Misidentification:** The incorrect identification of a known species as an unknown species. Misidentification is a common occurrence in cryptozoology and frequently stems from limited knowledge of known marine life, poor visibility conditions during sightings, or reliance on incomplete or unreliable evidence. Establishing a solid baseline understanding of known marine species is crucial to minimizing misidentification.

Investigative Methods and Data Analysis Terms:

- **Archival Research:** The examination of historical records, such as ship logs, nautical charts, and written accounts, to uncover clues about potential cryptids. Archival research plays a vital role in establishing the historical context of cryptid reports and identifying recurring themes or patterns that might shed light on their nature. Careful evaluation of source materials, considering the potential for bias or exaggeration, is crucial.
- **Data Analysis:** The systematic process of inspecting, cleaning, transforming, and modeling data with the goal of discovering useful information, informing conclusions, and supporting decision-making. In

cryptozoological investigations, data analysis might involve comparing sonar data, analyzing eyewitness accounts statistically, or mapping the geographical distribution of sightings. The use of appropriate statistical methods and visualization techniques is crucial for drawing valid conclusions from the collected data.

- **Geographic Information Systems (GIS):** A system designed to capture, store, manipulate, analyze, manage, and present all types of geographical data. GIS mapping plays a critical role in visualizing the geographical distribution of sightings, allowing for identification of patterns or correlations that might shed light on the nature of potential cryptids. Overlaying sighting data with environmental factors such as water depth, temperature, and salinity can reveal potential habitats or migratory routes.
- **Sonar Analysis:** The examination of sonar data to identify unusual acoustic signals or objects that may indicate the presence of unknown creatures. Sonar technology provides an effective means for detecting underwater objects, but the interpretation of sonar data requires expertise in both sonar technology and marine biology. False positives are common, and careful consideration of the environment and other factors is necessary to distinguish between natural phenomena and potential cryptids.
- **Underwater Drone Operation:** The use of remotely operated vehicles (ROVs) or autonomous un-

derwater vehicles (AUVs) to explore the underwater environment and gather visual and other data. Underwater drones are becoming increasingly valuable tools in cryptozoological investigations, allowing researchers to access previously inaccessible areas and gather high-quality visual evidence. However, the operation of underwater drones requires specialized training and expertise, and careful planning is crucial for maximizing the efficiency and safety of underwater missions.

This glossary serves as a starting point for understanding the specialized terminology used in the study of sea cryptids. As you progress in your investigations, you will encounter additional terms and concepts, and further research will be necessary. The responsible and ethical pursuit of cryptozoological research requires a solid foundation in both marine biology and investigative techniques. Remember that rigorous scientific methodology, balanced with open-mindedness, is crucial in this field.

This appendix provides a comprehensive index of key terms, concepts, and resources discussed throughout *An Investigator's Guide to Sea Cryptids.* Its purpose is to facilitate quick and easy navigation of the book's contents, allowing readers to readily locate specific information relevant to their research or interest. The index is structured alphabetically for ease of use and includes cross-referencing where appropriate, to connect related topics and concepts.

The index is divided into several sections to improve clarity and organization. The first section covers core concepts in marine biology, providing quick access to definitions and explanations of essential terms. This section is particularly crucial for those new to the field, offering a concise yet informative overview of relevant biological principles. The second section focuses on key terms specific to cryptozoology, including methodologies and analytical approaches. This section emphasizes the investigative aspects of studying potential cryptids, highlighting best practices and potential pitfalls. The third section lists investigative techniques and data analysis methods, providing a comprehensive overview of the practical tools and strategies involved in cryptozoological research. This section emphasizes the importance of rigorous data collection and analysis, promoting robust conclusions and minimizing bias. Finally, the fourth section lists resources, both primary and secondary, to assist readers in furthering their research and exploration of sea cryptids. This comprehensive index should serve as a valuable tool throughout your cryptozoological journey.

Index of Terms and Concepts:

A

Archival Research: See Investigative Methods. This encompasses the examination of historical records, including ship logs, nautical charts, whaling records, fishermen's accounts, and early scientific literature, to find potential evidence of unusual marine life. Critical evaluation of source material is crucial, assessing the credibility and potential biases of each source. Examples include examining reports of unusual sightings from 19th-century whaling logs, cross-referencing locations with known currents and seafloor topography, and analyzing historical maps for inconsistencies or annotations potentially hinting at unknown creatures.

Acoustic Signals: See Sonar Analysis. Analyzing the unique sound signatures of marine life, differentiating between known species and potential unknown organisms. This requires extensive knowledge of marine bioacoustics and the ability to interpret sonar data effectively. Advanced sonar technology can detect subtle differences in sound production, offering potential clues to identify unusual creatures.

Autonomous Underwater Vehicles (AUVs): See Underwater Drone Operation. These unmanned submersibles provide a powerful tool for deep-sea exploration, capable of collecting visual data, water samples, and other information in harsh environments. AUVs often carry multiple sensors, including cameras, sonar, and chemical sensors. The data gathered can reveal environmental factors alongside potential evidence of cryptids.

B

Baseline Understanding of Marine Life: Establishing a comprehensive knowledge of known marine fauna is fundamental before investigating potential cryptids. This involves understanding species identification, distribution, behaviors, and potential variations in appearance based on age, gender, and environmental factors. A strong foundation in marine biology minimizes misidentifications and allows for accurate assessment of unusual sightings. This includes not just visual identification but also understanding genetic variability, behavioral ecology and trophic interactions.

Benthic Zone: See Marine Biology Terms. The seabed ecosystem is extremely diverse, and understanding its habitats, including the variety of substrate types (sand, mud, rock) and the influence of hydrothermal vents or cold seeps, is critical. The distribution of benthic life influences where potential cryptids might reside.

Bioluminescence: See Marine Biology Terms. The study of bioluminescence patterns, including their intensity, frequency, and spectral characteristics, can help identify unknown species. The unique bioluminescent signatures of various creatures can serve as a powerful identification tool. Unusual bioluminescence should always be considered a significant lead.

C

Cephalopods: See Marine Biology Terms. Understanding the biology, behavior, and potential size variations within cephalopods is critical when examining reports of giant squid or other large cephalopod-like creatures. Knowledge of their intelligence, camouflage abilities, and deep-sea adaptations is crucial in assessing the plausibility of specific sightings.

Cold Seeps: See Hydrothermal Vents. These unique ecosystems support chemosynthetic life, vastly differing from the photosynthetic life in other ocean zones. They may harbor unique species not found elsewhere. Investigating cold seep communities is an important step in exploring potential habitats for cryptids.

Cryptozoology: See Cryptozoology Terms. The field's methodology emphasizes rigorous investigation using scientific methods, and a balance of open-mindedness and critical thinking. Ethical considerations are paramount, particularly concerning environmental impact and respect for marine ecosystems.

D

Data Analysis: See Investigative Methods. This includes statistical methods to analyze the frequency, location, and characteristics of reported sightings, along with visual analysis of photographs and videos. Careful data handling and quality control are essential. Techniques like GIS mapping are indispensable to understand spatial distribution and potential correlations with environmental variables.

Deep Scattering Layer (DSL): See Marine Biology Terms. Understanding its composition and variability helps differentiate potential cryptids from known species aggregations within the DSL. This involves analyzing sonar data in context with other environmental variables, such as depth, temperature, and time of day.

E

Eyewitness Testimony: See Cryptozoology Terms. While eyewitness accounts are a common source of information, their reliability depends on several factors, including lighting, distance, observer bias, and the observer's level of marine biology knowledge. Triangulating information from multiple independent witnesses strengthens the reliability of any given account. Detailed questioning and careful analysis are essential.

G

Geographic Information Systems (GIS): See Investigative Methods. GIS mapping is invaluable for visualizing the spatial distribution of sightings, identifying potential hotspots, and understanding correlations with environmental factors. Overlay techniques comparing sightings with bathymetry, temperature, salinity, or other datasets can unveil potential patterns.

Genetic Analysis: The use of genetic sequencing techniques to identify unknown organisms based on DNA or RNA samples. Environmental DNA (eDNA) analysis can detect the presence of organisms even without direct observation. This method provides strong scientific evidence and can help distinguish between known and unknown species.

H

Hydrothermal Vents: See Marine Biology Terms. These unique ecosystems are hotspots for biodiversity, potentially harboring yet-undiscovered organisms adapted to extreme conditions. Investigating hydrothermal vent communities can significantly advance understanding of deep-sea life and identify potential habitats for cryptids.

I

Investigative Methods: This encompasses a multifaceted approach, including archival research, sonar analysis, underwater drone operation, and genetic analysis. A structured, systematic approach is crucial, emphasizing careful data collection, rigorous analysis, and ethical considerations.

M

Marine Biology Terms: A core section providing explanations for terms fundamental to understanding ocean life. These are critical for evaluating the plausibility of any sea cryptid sighting. See the previous chapter for a detailed explanation.

Misidentification: See Cryptozoology Terms. A common challenge in cryptozoology. Thorough understanding of known marine life and careful analysis of evidence are essential to minimize misidentifications. Improved understanding of species variability and their reactions to different environments can lead to fewer misidentifications.

N

Nekton: See Marine Biology Terms. Understanding the distribution and behavior of known nektonic organisms is crucial for evaluating sightings of potential cryptids. This section helps differentiate between the movement patterns of known nekton and potential cryptic organisms. Consideration of swimming speed, depth range and migratory patterns aids in evaluation.

O

Oceanographic Data: Understanding ocean currents, temperature gradients, salinity, and other environmental factors is crucial in evaluating the plausibility of sightings and determining potential habitats for cryptids. Correlating sightings with oceanographic data helps identify potential migration patterns or preferred habitats of unknown species.

P

Pelagic Zone: See Marine Biology Terms. The vastness and relative unexplored nature of this zone makes it a likely habitat for undiscovered species. Understanding the different zones within the pelagic zone, based on depth and light penetration, is critical for assessing potential habitats and the movement patterns of potential cryptids.

Plankton: See Marine Biology Terms. Understanding the role and abundance of plankton helps in interpreting sonar data and accounting for possible misinterpretations. Large aggregations of plankton can create significant sonar reflections, potentially mistaken for large organisms.

R

Remotely Operated Vehicles (ROVs): See Underwater Drone Operation. ROVs, operated from a surface vessel, offer maneuverability and high-resolution imaging capabilities for close-up inspection of potentially interesting underwater features. ROVs can collect samples and capture high-quality video footage, providing valuable visual data.

S

Sonar Analysis: See Investigative Methods. Careful interpretation is crucial, accounting for factors such as water temperature, salinity, and seabed topography. Advanced signal processing techniques can help distinguish between natural phenomena and potential cryptids. Different types of sonar, with varying frequencies, might be required.

U

Underwater Drone Operation: See Investigative Methods. This involves the safe and efficient use of ROVs and AUVs. Careful planning and expertise are essential to obtain high-quality data without harming the environment or jeopardizing the safety of the equipment.

Z

Zooplankton: See Marine Biology Terms. Understanding the abundance, distribution, and potential for unusual aggregations of zooplankton helps distinguish them from potential cryptids in sonar data and visual observations. Unusual size or behavior should warrant further investigation.

Resources and Further Reading:

This section provides a starting point for further research into marine biology, cryptozoology, and investigative techniques. It is not exhaustive, and many other valuable resources exist. The goal is to supply a variety of sources, including primary research articles, authoritative textbooks, and reputable online databases.

Marine Biology:

- "Marine Biology" by Peter Castro and Michael Huber (Textbook) – Offers a comprehensive overview of marine ecosystems and organisms.
- "The World Ocean: An Introduction to Oceanography" by Paul R. Pinet (Textbook) – Provides a comprehensive foundation in oceanographic principles.
- "Oceanographic Literature Review" – A regularly updated online database providing access to thousands of scientific publications.
- *Journals:* Marine Ecology Progress Series, Limnology and Oceanography, Journal of Marine Research.

Cryptozoology:

- "The Encyclopedia of Cryptozoology" by Loren Coleman (Reference) – A comprehensive resource on cryptids worldwide.
- "Searching for Nessie: The Story of the Loch Ness Monster" by Roy Mackal (Case study) – A detailed analysis of a classic cryptozoological case.

- "Cryptozoology A to Z: The Encyclopedia of Loch Monsters, Sasquatch, Chupacabras, and Other Cryptic Creatures" by John E. Mack (Reference).
- Websites: Many reputable websites offer information on cryptids. However, critical evaluation of sources is essential, distinguishing between reliable and unreliable information. Be wary of sources with a lack of scientific rigor.

Investigative Techniques:

- "Forensic Science: An Introduction to Scientific and Investigative Techniques" by Stuart James (Textbook) – Provides a broad introduction to various investigative approaches.
- "Introduction to Geographic Information Systems" by Michael F. Goodchild (Textbook) – A textbook introducing GIS techniques for spatial data analysis.
- "Sonar for Oceanographers" by Robert J. Urick (Textbook) - Explains sonar technology and its applications.
- *Specialized Journals:* Journal of Field Ornithology, Journal of Mammalogy (for methodologies adaptable to marine environments).

This index serves as a preliminary guide. Continuous learning and exploration of new resources are crucial in the dynamic fields of marine biology and cryptozoology. Remember to always critically evaluate information and maintain ethical standards in your research.

www.ingramcontent.com/pod-product-compliance
Lightning Source LLC
La Vergne TN
LVHW052025200926
848481LV00005B/1589
9798349373350